D1302545

$\mathcal{S}$TAMPING THROUGH MATHEMATICS

Albrecht Dürer's enigmatic engraving *Melencolia I* appears on this miniature sheet from Mongolia. It features the brooding figure of Melancholy in reflective mood, holding a pair of compasses. Featured are a giant polyhedron, a sphere, an hourglass, and a 4 × 4 magic square in which the numbers in each row, column and diagonal add up to 34; the date of the engraving, 1514, appears in the bottom row.

$\mathcal{S}$TAMPING
THROUGH
MATHEMATICS

ROBIN J. WILSON

The Open University, UK

All science is either physics or stamp collecting

ERNEST RUTHERFORD

Springer

Robin J. Wilson
Department of Pure Mathematics
The Open University
Milton Keynes MK7 6AA
United Kingdom
r.j.wilson@open.ac.uk

Mathematics Subject Classification (2000): 01A05, 01Axx

Library of Congress Cataloging-in-Publication Data
Wilson, Robin J.
    Stamping through mathematics/Robin J. Wilson.
        p.   cm.
    Includes bibliographical references and index.
    ISBN 0-387-98949-8 (alk. paper)
    1. Mathematics—History.   2. Mathematics on postage stamps.   I. Title.

QA21.W39   2001
510'.9—dc21   00-052279

Printed on acid-free paper.

Production managed by Frank McGuckin; manufacturing supervised by Jacqui Ashri.
Typeset by Matrix Publishing Services, Inc., York, PA.
Printed and bound by Walsworth Publishing Company, Inc., Brookfield, MO.
Printed in the United States of America.

9 8 7 6 5 4 3 2 1

ISBN 0-387-98949-8   SPIN   10746624

Springer-Verlag New York Berlin Heidelberg
A member of BertelsmannSpringer Science-Business Media GmbH

Preface

There are many hundreds of postage stamps relating to mathematics, ranging from the earliest forms of counting to the modern computer age. In these pages you will meet many of the mathematicians who contributed to this story—influential figures such as Pythagoras, Archimedes, Newton and Einstein—and will learn about those areas, such as navigation, astronomy and art, whose study aided this development. Each topic appears on a double page, with a commentary on the left and enlarged stamp images on the right. A list of the featured stamps appears at the end of the book.

This book is written for anyone interested in mathematics and its applications. Although parts of it assume some knowledge of school or college level mathematics, I hope that much of it will be of interest to readers without this background. In particular, I hope that it will also attract a philatelic readership.

This is not a history of mathematics book in the conventional sense of the word. Several important mathematicians or topics are omitted, due to the absence of suitable stamps featuring them, whereas others may have assumed undue prominence because of the abundance of attractive images. Where appropriate I have felt free to let the stamps dictate the story.

Postage stamps are an attractive vehicle for presenting mathematics and its development. For some years I have successfully presented an illustrated lecture entitled *Stamping through mathematics* to school and college groups and to mathematical clubs and societies, and I am grateful to many people over the years for the useful comments they have made.

Since 1984 I have also contributed a regular 'Stamp Corner' to *The Mathematical Intelligencer*, and thank the publishers for permission to use material from these columns. Useful material for this book was also gleaned from *Philamath**, a regular news sheet for collectors of mathematical stamps. I am also very grateful to the Postal Authorities and individuals who have given permission to reproduce the copyrighted stamp images; a list of these appears in the *Acknowledgements* section at the end of the book.

Finally, many individuals have helped with suggestions, and I am particularly grateful to Marlow Anderson, June Barrow-Green, Joy Crispin-Wilson, Matthew Esplen, Florence Fasanelli, John Fauvel, Michael Ferguson, Raymond Flood, Paul Garcia, Helen Gardner, Caroline Grundy, Keith Hannabuss, Heiko Harborth, Roger Heath-Brown, Stephen Huggett, Victor Katz, Adrian Rice and Eleanor Robson for their support and advice. I am also very grateful to Tony Webb of the Open University for scanning the stamp images, and to Ina Lindemann, Joe Piliero and Jerry Lyons of Springer-Verlag, New York.

Robin Wilson
August 2000

* For information about *Philamath*, please contact: *Philamath*, 5615 Glenwood Road, Bethesda, MD 20817, USA

Contents

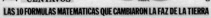

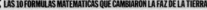
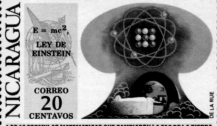

$\mathcal{S}$TAMPING
THROUGH
MATHEMATICS

The ten mathematical formulae that changed the face of the earth, as depicted by Nicaragua in 1971.

Early Mathematics

From earliest times people have needed to be able to count and measure the objects around them. Early methods of counting included forming stones into piles, cutting notches in sticks and **finger counting**. It is undoubtedly due to this last method that our familiar decimal number system emerged.

Early examples of mathematical writing appeared in Mesopotamia, between the rivers Tigris and Euphrates in present-day Iraq. A **Sumerian accounting tablet** from around 3000 BC features commodities such as barley; the three thumbnail indentations represent numbers. Their number system became a sexagesimal one, based on 60, which we still use in our measurement of time.

The Babylonians later imprinted their mathematics with a wedge-shaped stylus on damp clay which was then baked in the sun. Hundreds of these cuneiform tablets from 1900 to 1600 BC have survived, and show a good understanding of arithmetic (including a very accurate value for $\sqrt{2}$), algebra (the solution of linear and quadratic equations) and geometry (the calculation of areas and volumes). There is also a tablet indicating a detailed knowledge of Pythagorean triples (numbers a, b, c satisfying $a^2 + b^2 = c^2$) a thousand years before Pythagoras; one of the triples is 12,709, 13,500, 18,541—a remarkable achievement for the time. The Babylonians also studied astronomy and were able to predict eclipses; in 164 BC they observed the comet now known as **Halley's comet** (see page 60)—not in 2349 BC as stated on the stamp opposite.

Geometrical alignments of stones have been found in several places. Celebrated examples include the circular pattern of megaliths at **Stonehenge** and the linear arrangements at **Carnac** in Brittany. Although their exact purpose is unknown, it is likely that their construction had religious significance and was designed to demonstrate astronomical events such as sunrise on midsummer's day.

Interest in geometrical patterns can also be seen in cave drawings. Attractive examples of early **geometrical cave art** have been found in the Chuquisaca region of Bolivia.

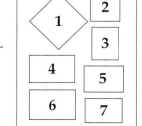

1. *finger counting*
2. *finger counting*
3. *geometrical cave art*
4. *Sumerian accounting tablet*
5. *Babylonian tablet and comet*
6. *Stonehenge*
7. *Carnac*

HISTORY OF WRITING : PICTOGRAPHIC SCRIPT SUMERIAN TABLET

Egypt

The main achievements of early Egyptian mathematicians involved the practical skill of measurement. The oldest of the Egyptian pyramids is **King Djoser's step pyramid** in nearby Saqqara, built in horizontal layers and dating from about 2700 BC. It was supposedly designed by **Imhotep**, the celebrated court physician, Grand Vizier and architect.

The magnificent **pyramids of Gizeh** (or Giza) date from about 2600 BC and attest to the Egyptians' extremely accurate measuring ability. In particular, the Great Pyramid of Cheops has a square base whose sides of length 230-metre agree to less than 0.01%. Constructed from more than two million blocks averaging over 2 tonnes in weight, the pyramid is 146 metres high and contains an intricate arrangement of internal chambers and passageways.

Our knowledge of later Egyptian mathematics is scanty, deriving mainly from two fragile primary sources, the *Moscow papyrus* (c.1850 BC) and the *Rhind papyrus* (c.1650 BC). These papyri include tables of fractions and several dozen solved problems in arithmetic and geometry, probably designed for the teaching of scribes and accountants. These problems range from division problems involving the sharing of a number of loaves in specified proportions to geometrical problems on the area of a triangle of land and the volume of a cylindrical granary of given diameter and height; the solution of the latter problem gives rise to a value of π of $256/81$ (about 3.16). The Egyptians used a decimal system for counting, but their fractions were mainly 'unit fractions' of the form $1/n$; more complicated fractions were written in terms of these—for example,

$$\frac{13}{21} = \frac{1}{3} + \frac{1}{4} + \frac{1}{28}.$$

Prominent among other Egyptians interested in mathematics was **Amenhotep** (see page 75), a high official during the reign of Amenhotep III (c.1400 BC). During the Ptolemaic era 1000 years later, his name came to be associated with the ibis-headed **Thoth**, the Egyptian god of reckoning.

1. *King Djoser's pyramid*   2. *Gizeh and pyramids*
3. *pyramids of Gizeh*   4. *Egyptian accountants*
5. *Egyptian papyrus*   6. *Imhotep*
7. *Thoth*

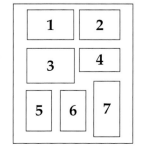

Greek Geometry

Starting from around 600 BC, mathematics and astronomy flourished for over one thousand years throughout the Greek-speaking world of the eastern Mediterranean Sea. During this time, the Greeks developed the concept of deductive logical reasoning that became the hallmark of much of their work, especially in the area of geometry. Some of their achievements are described in the next few pages.

The earliest known Greek mathematician is **Thales of Miletus** (c.624–547 BC) who, according to legend, brought geometry to Greece from Egypt. He predicted a solar eclipse in 585 BC and showed how rubbing with a stone can produce electricity in feathers. In geometry he investigated the congruence of triangles, applying it to navigation at sea, and is credited with proving that the base angles of an isosceles triangle are equal and that a circle is bisected by any diameter.

Pythagoras (c.572–497 BC) was a semi-legendary figure. Born on the Aegean island of Samos, he later emigrated to the Greek seaport of Crotona, now in Italy, where he founded the Pythagorean school. This closely-knit brotherhood was formed, according to later writers, to further the study of mathematics, philosophy and the natural sciences. The Pythagoreans believed that 'All is number', and there was a particular emphasis on the 'mathematical arts' of arithmetic, geometry, astronomy, and music. Pythagoras is one of the Greek scholars depicted in Raphael's Vatican fresco *The School of Athens* (c.1509).

It is not known who first proved **Pythagoras' theorem**, that the area of the square on the hypotenuse of a right-angled triangle is the sum of the areas of the squares on the other two sides, but Pythagorean triples were already known to the Babylonians (see page 2).

Democritus (c.400–370 BC) was also interested in geometry, investigating the properties of pyramids and cones by splitting them into 'indivisible' sections by planes parallel to the base. He is primarily known for first proposing the view that all matter consists of small indivisible particles, called atoms.

1. $3^2 + 4^2 = 5^2$
2. *Greek coin showing Pythagoras*
3. *Pythagoras' theorem*
4. *Pythagoras' theorem*
5. *Thales of Miletus*
6. *Pythagoras ('School of Athens')*
7. *Democritus*

1	2
3	4
5	6
7	

Plato's Academy

From about 500 to 300 BC, Athens became the most important intellectual centre in Greece, numbering among its scholars Plato and Aristotle. Although neither is remembered primarily as a mathematician, both helped to set the stage for the 'golden age of Greek mathematics' in Alexandria.

The Acropolis, the 'highest city of ancient kings' was devastated by a Persian invasion in 480 BC. Inspired by Pericles, the city was rebuilt and the magnificent **Parthenon** was added, being completed in 432 BC. Constructed on mathematical principles, it is surrounded on all sides by towering columns of white marble.

Around 387 BC, **Plato** (c.427–347 BC) founded his school in a part of Athens called Academy. Here he wrote and directed studies, and the Academy soon became the focal point for mathematical study and philosophical research. Over the entrance appeared the inscription: 'Let no-one ignorant of geometry enter here'.

Plato believed that the study of mathematics and philosophy provided the finest training for those who were to hold positions of responsibility in the state. In his *Republic* he discussed the Pythagoreans' mathematical arts of arithmetic, plane and solid geometry, astronomy and music, explaining their nature and justifying their importance for the 'philosopher-ruler'. His *Timaeus* includes a discussion of the five regular solids—tetrahedron, cube, octahedron, dodecahedron and icosahedron.

Aristotle (384–322 BC) became a student at the Academy at the age of 17 and stayed there for twenty years until Plato's death. He was fascinated by logical questions and systematised the study of logic and deductive reasoning. In particular, he mentioned a proof that $\sqrt{2}$ cannot be written in rational form a/b, where a and b are integers, and he discussed syllogisms such as:

'all men are mortal; Socrates is a man; thus Socrates is mortal'.

In Raphael's fresco *The School of Athens* (see page 6), Plato and Aristotle are pictured on the steps of the Academy. Plato is holding a copy of his *Timaeus* and Aristotle is carrying his *Ethics*.

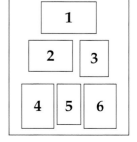

1. *Parthenon*
3. *bust of Plato*
5. *Byzantine fresco of Aristotle*
6. *Greek map and base of statue*

2. *Roman statue of Aristotle*
4. *Plato and Aristotle*
   ('School of Athens')

Euclid and Archimedes

Around 300 BC, with the rise to power of Ptolemy I, mathematical activity moved to the Egyptian part of the Greek empire. In **Alexandria** Ptolemy founded a university that became the intellectual centre for Greek scholarship for over 800 years. Ptolemy also started its famous library, which eventually held over half-a-million manuscripts. The celebrated Pharos lighthouse at Alexandria, seen on the stamp opposite, was one of the seven wonders of the ancient world.

The first important mathematician associated with Alexandria was **Euclid** (c.300 BC), who wrote on optics and conics, but is mainly remembered for his *Elements*. The most influential and widely read mathematical book of all time, the *Elements*, is a compilation of results known at the time, and consists of thirteen books on plane and solid geometry, number theory, and the theory of proportion. A model of deductive reasoning, it starts from initial axioms and postulates and uses rules of deduction to derive each new proposition in a logical and systematic order.

Archimedes (c.287–212 BC), a native of Syracuse on the island of Sicily, was one of the greatest mathematicians who ever lived. In geometry he calculated the surface areas and volumes of various solids, such as the sphere and cylinder, and listed the thirteen semi-regular solids whose faces are regular but not all of the same shape. By considering 96-sided polygons that approximate a circle, Archimedes proved that π lies between $3\tfrac{10}{71}$ and $3\tfrac{10}{70}$ ($= \tfrac{22}{7}$), and he also investigated the 'Archimedean spiral', now usually written with polar equation $r = k\theta$.

In applied mathematics he made outstanding contributions to both mechanics and statics. In mechanics he found the 'law of moments' for a balance with weights attached, devised ingenious mechanical contrivances for the defence of Syracuse, and is credited with inventing the **Archimedean screw** for raising water from a river. In statics he observed that the weight of an object immersed in water is reduced by an amount equal to the weight of water displaced—now called 'Archimedes' principle'—and used this to test the purity of King Hiero's gold crown. On discovering his principle he supposedly jumped out of his bath and ran naked down the street shouting 'Eureka!' (I have found it!).

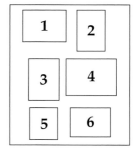

1. *Alexandria*   4. *Euclid and his pupils ('School of Athens')*
2. *Euclid*   5. *Archimedes (Ribera portrait)*
3. *Archimedes*   6. *Archimedes and screw*

Greek Astronomy

The heavens were studied by a number of Greek scholars—in particular, Eudoxus, Aristarchus, Hipparchus and Ptolemy.

The mathematician and astronomer **Eudoxus of Cnidus** (408–355 BC) studied at Plato's Academy and is often credited with developing the theory behind Books V (on proportion) and XII (on the 'method of exhaustion') of Euclid's *Elements*. In astronomy he advanced the hypothesis that the sun, moon and planets move around the earth on rotating concentric spheres, a hypothesis later adopted in modified form by Aristotle.

Aristarchus of Samos (c.310–230 BC) advanced an alternative hypothesis—that 'the fixed stars and the sun remain unmoved and that the earth revolves about the sun in the circumference of a circle'. Aristarchus thus anticipated by 1800 years the revolutionary work of Nicolaus Copernicus (see page 44), but his hypothesis found few adherents among his contemporaries.

The first trigonometrical approach to astronomy was provided by **Hipparchus of Bithynia** (190–120 BC), possibly the greatest astronomical observer of antiquity. Sometimes called 'the father of trigonometry', he discovered the precession of the equinoxes and constructed a 'table of chords' yielding the sines of angles. He also introduced a coordinate system for the stars and constructed the first known star catalogue.

The earth-centred hypothesis was developed by Claudius Ptolemy of Alexandria (c.100–178 AD), giving rise to the **Ptolemaic system**, shown opposite on a block of stamps honouring Copernicus. Ptolemy wrote a definitive 13-volume work on astronomy, usually known by its later Arabic name *Almagest* ('the greatest'). It contained a mathematical description of the motion of the sun, moon and planets, and included a table of chords listing the sines of angles from 0° to 180° in steps of ½°.

Ptolemy also published a standard work on map-making called *Geographia*, in which he discussed various types of map projection and gave the latitude and longitude of 8000 places in the known world. His maps were used by navigators for many centuries.

1. *Ptolemaic planetary system*
2. *Hipparchus*
3. *Aristarchus' planetary system*
4. *Aristarchus' theory and diagram*
5. *Eudoxus' solar system*

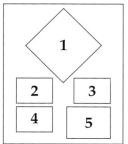

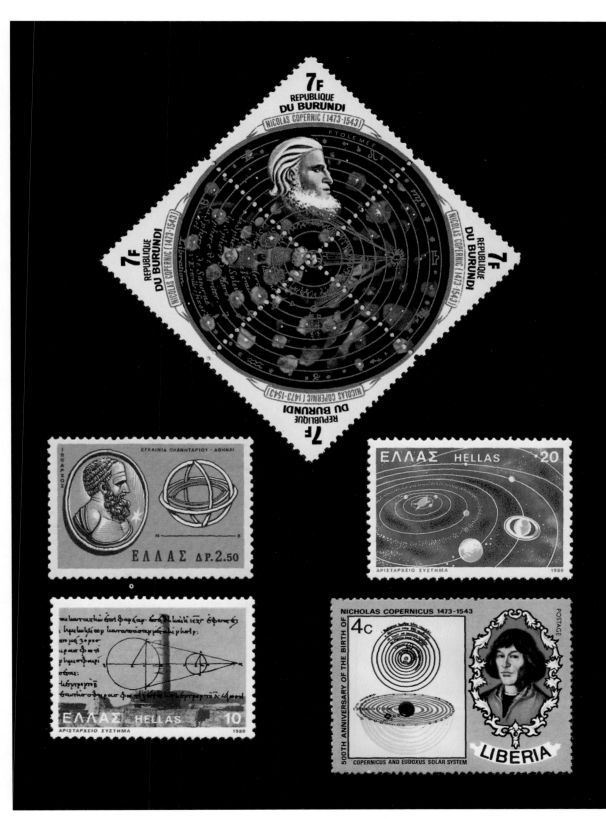

Mathematical Recreations

Games have a universal appeal and have been played since earliest times. Many games require great skill and ingenuity and have been subjected to much mathematical analysis. Go and chess are discussed on page 34; here we concentrate on some less familiar games.

Senet is an early form of backgammon and may date back to 3000 BC. A celebrated Egyptian example from 1350 BC was found in the tomb of Tutankhamen. It was played by two players on a 3×10 board with lion-shaped pieces. The African game of **eklan** is also played by two players. It consists of a board with 24 holes, arranged in concentric squares, into which sticks are inserted according to specified rules.

Mancala is a count-and-capture board game, usually played with counters (pebbles or beans) by two players. The board has a number of indentations, generally arranged in two rows, with a larger compartment (the 'mancala') at the end on each player's right. The players place counters into the indentations and move them according to specified rules, in order to capture their opponent's counters. Variations on mancala have appeared in many parts of Africa and southern Asia. In Indonesia it is known as **dakon** (or tjongkak), while an early form involving four rows of indentations was played in southern Africa under the name **morabaraba**.

Baghchal, found in Nepal, is a form of draughts or checkers. The earliest game of this type seems to be alquerque or el-quirkat, a game found in the Middle East around 1400 BC. It is played on a 5×5 board by two players, who move their pieces along the lines and capture their opponent's pieces by 'jumping over' them.

Mazes and labyrinths have existed for many thousands of years. In ancient Crete, King Minos ruled over the intricate labyrinth of Knossos where Greek youths were regularly sacrificed to a fierce minotaur that was eventually slain by Theseus. Such labyrinths sometimes appeared on Cretan coins; the stamp opposite shows a **seven-ring labyrinth** on a coin dating from about 450 BC.

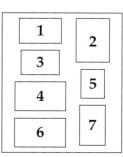

1. *senet board*
2. *baghchal board*
3. *morabaraba*
4. *playing eklan*
5. *dakon (tjongkak)*
6. *eklan board*
7. *Cretan labyrinth*

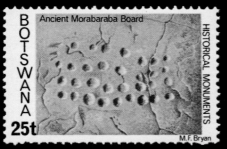

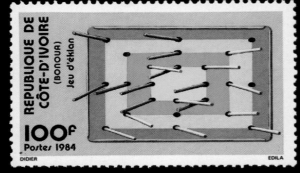

China

Most ancient Chinese mathematics was written on bamboo or paper which has perished with time. However, an outstanding work that survives, possibly from 200 BC, is the *Jiuzhang suanshu* [Chiu-chang suan-shu] (Nine chapters on the mathematical art), which contains the calculation of areas and volumes, the evaluation of square and cube roots, and the systematic solution of simultaneous equations.

Several Chinese mathematicians devoted their attention to evaluating π. **Zhang Heng** [Chang Heng] (78–139 AD), inventor of the seismograph for measuring earthquake intensity, proposed the value $\sqrt{10}$ (about 3.16), a value also found by Indian mathematicians a couple of centuries earlier. **Liu Hui** (c.260 AD), while revising the *Jiuzhang suanshu*, calculated the areas of regular polygons with 96 and 192 sides and deduced that π lies between 3.1410 and 3.1427. Most remarkable was the work of **Zu Changzhi** [Tsu Ch'ung-Chih] (429–500), who calculated the areas of regular polygons with 12,288 and 24,576 sides and deduced that π lies between 3.1415926 and 3.1415927. Zu Changzhi also obtained the approximation $355/113$, which is correct to six decimal places; such accuracy was not obtained in the West for another thousand years.

The thirteenth and fourteenth centuries saw Chinese contributions to algebra and the numerical solution of equations. The **arithmetical triangle** of binomial coefficients, now usually called 'Pascal's triangle', appeared in a Chinese text of 1303. A notable figure of the time was **Guo Shoujing** [Kuo Shou-Ching] (1231–1316), who worked on calendar construction, astronomy and spherical trigonometry.

Various Chinese measuring instruments have survived, including a **distance-measuring drum cart** from around 300 AD and a 1437 **armillary sphere**, an astronomical device for representing the great circles of the heavens. The **abacus** has appeared in different forms around the world, originally as a sand tray with pebbles; the Chinese version consists of a frame with beads, called a 'suan pan'.

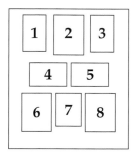

1. *Zu Changzhi*   2. *Liu Hui's evaluation of π*
3. *Zhang Heng*   4. *distance-measuring cart*
5. *armillary sphere*   6. *Chinese abacus*
7. *Guo Shoujing*   8. *arithmetical triangle*

A REFIND VALUE OF π 3RD CENTURY AD

弧 田 圓

圓周率

LUI HUI'S NINE CHAPTER ON THE MATHEMATICS AD 264

FEDERATED STATE OF
MICRONESIA   33¢

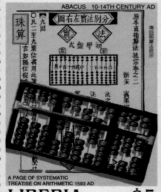

ABACUS 10-14TH CENTURY AD

珠算

A PAGE OF SYSTEMATIC
TREATISE ON ARITHMETIC 1593 AD

LIBERIA $5

'PASCAL' TRIANGLE 12TH CENTURY AD

A DIAGRAM OF 'FROM 'PRECIOUS MIRROR OF THE FOUR ELE-
MENT' PUBLISHED IN 1303

LIBERIA $5

India

Around 250 BC King Ashoka, ruler of most of India, became the first Buddhist monarch. His conversion was celebrated by the construction of many pillars carved with his edicts. These **Ashoka columns** contain the earliest known appearance of what would eventually become our Hindu-Arabic numerals. The Nepalese stamp opposite shows the Ashoka column in Lumbini, the birthplace of Buddha.

Unlike the complicated Roman numerals, and the Greek decimal system in which different symbols were used for 1, 2, . . . , 9, 10, 20, . . . , 90, 100, 200, . . . , 900, etc., the Hindu number system uses the same ten digits throughout, but in a place-value system where the position of each digit indicates its value. This enables calculations to be carried out column by column.

Indian mathematics can be traced back to around 600 BC. A number of **Vedic manuscripts** contain early work on arithmetic, permutations and combinations, the theory of numbers and the extraction of square roots.

The two most outstanding Indian mathematicians of the first millennium AD were Aryabhata (b. 476) and Brahmagupta (b. 598). Aryabhata gave the first systematic treatment of 'Diophantine equations'—algebraic equations for which we seek solutions in integers. He also presented formulas for the sum of the natural numbers and of their squares and cubes and obtained the value 3.1416 for π. The first Indian satellite was named **Aryabhata** in his honour, and appears on the stamp opposite.

Brahmagupta discussed the use of zero (another Indian invention) and negative numbers, and described a general method for solving quadratic equations. He also solved some quadratic Diophantine equations such as $92x^2 + 1 = y^2$ (now known as 'Pell's equation'), for which he obtained the integer solution $x = 120$, $y = 1151$. Around this time the game of chess was invented in India (see page 34); the stamp opposite features an Indian chess piece from the eighteenth century.

In later years Indian mathematicians and astronomers became interested in practical astronomy and built magnificent observatories such as the **Jantar Mantar** in Jaipur.

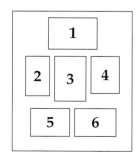

1. Vedic manuscript   2. Ashoka column capital
3. Indian chess piece   4. Ashoka column, Lumbini
5. Aryabhata satellite   6. Jantar Mantar

Mayans and Incas

One of the most interesting counting systems is that of the Mayans of Central America (Mexico, Guatemala, Belize, El Salvador and Honduras), between 300 and 1000 AD. Based on the numbers 20 and 18, it was a place-value system composed of the symbols • for 1, —— for 5 (combined to give numbers up to 19) and ◡ for 0; for example,

$$• \ •• \ ◡ \ \underline{••••} \ \text{ means } (1 \times 7200) + (7 \times 360) + (0 \times 20) + 19 = 9739.$$

Each number was also written in pictorial form, the picture representing the head of a man, bird, animal or deity.

Most of their mathematical calculations involved the measurement of time. They used two calendars: a 260-day ritual calendar with 13 cycles of 20 days, and a 365-day calendar with 18 months of 20 days and five extra days. Combining these calendars gave a long 'calendar-round' of 18,980 days (= 52 calendar years).

Our knowledge of the Mayans' counting system and of their calendars is derived mainly from hieroglyphic inscriptions on carved pillars (stelae), writings on the walls of caves and ruins, and a handful of painted manuscripts. The most notable of these manuscripts is the beautiful **Dresden codex**, dating from about 1200 AD. It is painted in colour on a long strip of glazed fig-tree bark and contains many examples of Mayan numbers.

Around the year 1500 the Incas of Peru, two thousand miles further south, invented the **quipu** for recording and conveying numbers and other statistical information. It consists of a main cord to which many thinner knotted cords of various colours are attached; the size and position of each knot correspond to a different number in a decimal system, and the colours convey different types of information. Quipus were used for accounting purposes and for recording other types of numerical data. The information was carried around the region by teams of **Inca messengers**, trained runners who could cover large distances in a day.

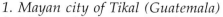

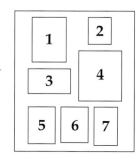

1. *Mayan city of Tikal (Guatemala)*   2. *Mayan calendar stone*
3. *Mayan observatory (Mexico)*   4. *Dresden codex*
5. *Inca messenger*   6. *Peruvian quipu*
7. *Dresden codex*

Early Islamic Mathematics

The period from 750 to 1400 was an important time for the development of mathematics. United by their new religion, Islamic scholars seized on the available Greek and Roman writings from the west and Hindu writings from the east and developed them considerably. Their achievements are described in the next few pages.

Some of our mathematical language dates from the Islamic period. The word 'algorithm' (a routine step-by-step procedure for solving a problem) derives from the name of the Persian mathematician **Muhammad ibn Musa al-Khwarizmi** (c.780–850), who lived in Baghdad and wrote influential works on arithmetic and algebra. His *Arithmetic* is important for introducing the Hindu decimal place-value system to the Islamic world, and the title of his algebra book, *Kitab al-jabr wal-muqabala* gives us the term 'algebra'; the word 'al-jabr' refers to the operation of adding a positive quantity to eliminate a negative one.

The first outstanding philosopher of the period was **al-Kindi** (d. c.870), who produced over two hundred works on subjects ranging from Euclid's geometry and Indian arithmetic to cooking. Another was the brilliant Islamic scholar **al-Farabi** (c.878–950), whose mathematical writings included an influential commentary on Euclid's *Elements*.

Al-Biruni (973–1055) was an outstanding intellectual figure who contributed over one hundred works, primarily on arithmetic and geometry, astronomy, geography and the calendar. Writing on trigonometry, he was one of the first mathematicians to investigate the tangent, cotangent, secant and cosecant functions.

The most celebrated of all the Arabian philosopher-scientists, primarily known for his treatises on medicine, was **ibn Sinah** (980–1037), usually known in the west as Avicenna. He contributed to arithmetic and number theory, produced a celebrated Arabic summary of Euclid's *Elements*, and applied his mathematical knowledge to various problems from physics and astronomy.

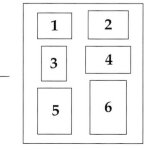

1. *Arabic science*   2. *al-Biruni*
3. *al-Khwarizmi*   4. *ibn Sinah (Avicenna)*
5. *al-Kindi*   6. *al-Farabi*

The Middle Islamic Period

During the eighth and ninth centuries, the Islamic world spread along the northern coast of Africa and up through southern Spain and Italy. The next three centuries were to be fertile years for the development of mathematics and astronomy. The Islamic astronomers began to use the astronomical and navigational instruments of the day. These included the planisphere and the astrolabe, used for observing the positions and altitudes of celestial bodies and for telling the time of day.

Influential among the works of the time, especially when translated into Latin during the Renaissance, were those of **ibn al-Haitham** (965–1039). Known in the west as Alhazen, he was a geometer whose main contributions were to the study of optics. A celebrated problem is 'Alhazen's problem', which asks: at which point on a spherical mirror is light from a given point source reflected into the eye of a given observer? An equivalent formulation is: at which point on the cushion of a circular billiard table must a cue ball be aimed so as to hit a given target ball?

Omar Khayyam (1048–1131) was a mathematician and poet who wrote on the binomial theorem and on geometry. In algebra he presented the first systematic classification of cubic equations and a discussion of their solution; such equations were not to be solved in general until the sixteenth century. He is known in the west mainly for his collection of poems known as the *Rubaiyat*.

Omar Khayyam also publicly criticised an attempted proof by ibn al-Haitham of Euclid's so-called 'parallel postulate' (see page 70). A later unsuccessful attempt to prove this postulate was given by the distinguished Persian mathematician **Nasir al-Din al-Tusi** (1201–1274). Al-Tusi constructed the first modern astronomical observatory and wrote influential treatises on astronomy, logic, theology and ethics. His several contributions to trigonometry included the sine rule for triangles: if *a*, *b* and *c* are the sides of a triangle, opposite to the angles *A*, *B* and *C*, then

$$a/\sin A = b/\sin B = c/\sin C.$$

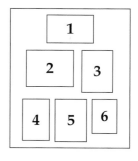

1. *al-Haitham's optics*
2. *Omar Khayyam: 'myself when young'*
3. *Nasir al-Din al-Tusi*
4. *Istanbul astronomers*
5. *Persian planisphere*
6. *Omar Khayyam*

Late Islamic Mathematics

During the Middle Ages, Córdoba became the scientific capital of Europe. Islamic decorative art and architecture spread throughout southern Spain and Portugal; celebrated examples include the magnificent arches in the **Córdoba Mezquita** (mosque) and the variety of geometrical tiling patterns in the Alhambra in Granada.

The Spanish geometer and astronomer **al-Zarqali** (d. 1100) lived in Toledo and Córdoba and produced several important works, including a set of trigonometrical tables, chiefly for use in astronomy. He was an instrument maker who constructed a number of astrolabes.

The Jewish scholar Moses ben Maimon, or **Maimonides** (1135–1204), was also born in Córdoba. Religious persecution forced his family to move to Cairo where he practised medicine, becoming personal physician to Saladin, Sultan of Egypt. A mathematician, astronomer and philosopher, he wrote on the calendar, the moon and combinatorial problems, and asserted that π is irrational. Another Córdoban was the influential commentator **Muhammad ibn Rushd** (1126–1198), known in the west as Averroës, who translated the works of Aristotle into Arabic and wrote a treatise specifying which parts of Euclid's *Elements* were needed for the study of Ptolemy's *Almagest*.

By the fifteenth century Samarkand in central Asia had become one of the greatest centres of civilisation. The Persian mathematician and astronomer **Jamshid al-Kashani** (or al-Kashi) (d. 1429) made extensive calculations with decimal fractions and established a notation for them, using a vertical line to separate the integer and fractional parts. A prodigious calculator, he determined π to 16 decimal places and obtained a very precise value for the sine of $1°$, from which many other trigonometrical values can be determined.

Al-Kashani's patron was the Turkish astronomer **Ulugh Beg** (1394–1449), whose observatory contained a special sextant, the largest of its type in the world. Using al-Kashani's value for sin $1°$, Ulugh Beg constructed extensive tables for the sine and tangent of every angle for each minute of arc to five sexagesimal places.

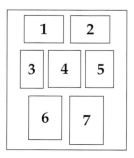

1. *Córdoba Mezquita*
2. *al-Zarqali and astrolabe*
3. *Maimonides*
4. *Arabic tile*
5. *ibn Rushd (Averroës)*
6. *al-Kashani*
7. *Ulugh Beg's observatory*

The Middle Ages

The period from 500 to 1000 in Europe is known as the Dark Ages. The legacy of the ancient world was almost forgotten, schooling became infrequent, and the general level of culture remained low. Mathematical activity was generally sparse, but included some writings on the calendar and on finger reckoning by the Venerable Bede (c.673–735), and the influential *Problems for the quickening of the mind* by Alcuin of York (735–804), educational adviser to Charlemagne.

Revival of interest in mathematics began with **Gerbert of Aurillac** (938–1003), who trained in Catalonia and was probably the first to introduce the Hindu-Arabic numerals to Christian Europe, using an abacus that he had designed for the purpose; he was crowned Pope Sylvester II in 999. Hindu-Arabic methods of calculation were also used by Fibonacci (Leonardo of Pisa) in his *Liber abaci* [Book of calculation] of 1202. This celebrated book contained many problems in arithmetic and algebra, including the celebrated problem of the rabbits that leads to the 'Fibonacci sequence' 1, 1, 2, 3, 5, 8, 13, . . . , in which each term after the first two is the sum of the previous pair.

Several other distinguished scholars studied mathematics around this time. These included **Albertus Magnus** (c.1193–1280), who introduced the works of Aristotle to European audiences, and **Geoffrey Chaucer** (1342–1400), author of the *Canterbury tales*, who wrote a treatise on the astrolabe, one of the earliest science books to be written in English. John of Gmunden (1384–1442) also discussed astronomical instruments, including astrolabes, quadrants, and sundials; the attractive **clock of Imms** (1555) is based on his designs. The cardinal and scholar **Nicholas of Cusa** (1401–1464) wrote several mathematical tracts and attempted the classical problems of trisecting an angle and squaring the circle; he also invented concave lens spectacles.

The Catalan mystic **Ramon Lull** (c.1232–1316) believed that all knowledge can be obtained as mathematical combinations of a fixed number of 'divine attributes'. Lull's ideas spread through Europe, and influenced later mathematicians such as Leibniz.

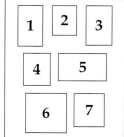

1. *Ramon Lull*
2. *Nicholas of Cusa*
3. *Gerbert of Aurillac*
4. *Albertus Magnus*
5. *Albertus Magnus*
6. *Geoffrey Chaucer*
7. *clock of Imms*

The Growth of Learning

The renaissance in mathematical learning during the Middle Ages was largely due to three factors: the translation of Arabic classical texts into Latin during the twelfth and thirteenth centuries, the establishment of the earliest European universities, and the invention of printing. The first of these made the works of Euclid, Archimedes and other Greek writers available to European scholars, the second enabled groups of like-minded scholars to meet and discourse on matters of common interest, while the last enabled scholarly works to be available at modest cost to the general populace.

The first European university was founded in Bologna in 1088, and Paris and Oxford followed shortly after. The curriculum was in two parts. The first part, studied by those aspiring to a Bachelor's degree, was based on the ancient 'trivium' of grammar, rhetoric and logic (usually Aristotelian). The second part, leading to a Master's degree, was based on the 'quadrivium', the Greek mathematical arts of arithmetic, geometry, astronomy and music; the works studied would have included Euclid's *Elements* and Ptolemy's *Almagest*.

Johann Gutenberg's invention of the printing press (around 1440) enabled classic mathematical works to be widely available for the first time. At first the new books were printed in Latin for the scholar, but increasingly vernacular works began to appear at a price accessible to all; these included texts in arithmetic, algebra and geometry, as well as practical works designed to prepare young men for a commercial career.

Important among the new printed texts was the 1494 *Summa de arithmetica, geometrica, proportioni et proportionalita* of **Luca Pacioli** (1445–1517), a 600-page compilation of the mathematics known at the time; it included the first published account of double-entry bookkeeping. In Germany the most influential of the commercial arithmetics was by **Adam Riese** (c.1489–1559); it proved so reputable that the phrase 'nach Adam Riese' [after Adam Riese] came to indicate a correct calculation.

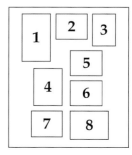

1. University of Bologna
2. university scholars
3. printing press
4. Bologna students
5. arithmetic and geometry
6. astronomy and music
7. Adam Riese
8. Luca Pacioli

Renaissance Art

Connections between mathematics and the visual arts have been apparent since earliest times. Some early geometrical cave art was featured on page 3, and many peoples have incorporated mathematical patterns into the designs of their pots and vases and in their weaving and basketry. The Romans frequently used geometrical decorations in their mosaics.

One notable feature of Renaissance art was that, for the first time, painters became interested in depicting three-dimensional objects realistically, giving visual depth to their works. This soon led to the formal study of geometrical perspective.

The first person to investigate perspective seriously was the artisan-engineer **Filipo Brunelleschi** (1377–1446), who designed the self-supporting octagonal cupola of the cathedral in Florence. Brunelleschi's ideas were developed by his friend **Leon Battista Alberti** (1404–1472), who presented mathematical rules for correct perspective painting and stated in his *Della pittura* [On painting] that 'the first duty of a painter is to know geometry'.

Piero della Francesca (1415–1492) found a perspective grid useful for his investigations into solid geometry, and wrote *De prospectiva pingendi* [On the perspective of painting] and *Libellus de quinque corporibus regularibus* [Book on the five regular solids]. The picture on the stamps opposite, his painting *Madonna and child with saints* (1472), is in perfect mathematical perspective.

The other title on these stamps is *De divina proportione* [On divine proportion] (1509) by Piero's friend Luca Pacioli (see page 30); Pacioli is the monk depicted second from the right. The woodcuts of polyhedra in this book were by Pacioli's friend and student **Leonardo da Vinci** (1452–1519), who explored perspective more deeply than any other Renaissance painter. In his *Trattato della pittura* [Treatise on painting] da Vinci warns 'Let no one who is not a mathematician read my work'.

Albrecht Dürer (1471–1528) was a celebrated German artist and engraver who learned perspective from the Italians and introduced it to Germany. His famous engravings, such as *Melencolia I* (see frontispiece) and *St Jerome in his study*, show his effective use of perspective.

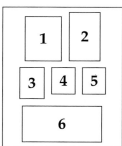

1. *Dürer's 'St Jerome in his study'*
2. *Roman mosaic*
3. *Leonardo da Vinci*
4. *Filipo Brunelleschi*
5. *Leon Battista Alberti*
6. *Piero della Francesca's 'Madonna and child with saints'*

Go and Chess

On page 14 we illustrated some board games from around the world. Two of the most celebrated games, Go and chess, have been extensively subjected to mathematical analysis and have featured on many stamps.

The ancient game of **Go**, formerly 'wei-chi' or 'weiqi', was created in China at least three thousand years ago and was introduced to Japan and Korea during the Tang Dynasty (618–907), before eventually finding its way to Europe. It is now usually played on a 19×19 board by two players who alternately place black and white stones on the intersections of the squares. The object of the game is to capture one's opponent's stones by surrounding them, and thereby control large areas of the board. Two **Go formations**, 'China vogue' (black) and 'linked stars' (white), are depicted on the Chinese stamp opposite, and **Choe Un**, a seven-year-old competitor in the World Go Championships, appears on the stamp from North Korea.

The game of **chess** probably originated in the sixth century in India, where it developed from a game called 'chaturanga'. It quickly spread to Persia, where it was known as 'shatranj', and from there to the Arabic world. During the eighth and ninth centuries the Moors took the game to Spain, and thence to the rest of Europe where it had become widely established by the eleventh century.

The oldest and most celebrated European work on chess is the beautifully illustrated *Book of chess, dice and boards*, commissioned in 1283 by King Alfonso the Wise of Castile and León; a diagram from this book appears on the stamp from Yemen opposite. Other chess books followed—notably, William Caxton's *The game and playe of chesse*, which appeared in 1476. An Italian book of the late fifteenth century illustrates a **royal chess party** that took place in Florence in 1493.

Variations on the game of chess have been played on boards of various sizes, such as those of sizes 8×9, 8×12, 9×9 and 12×12. It was not until the seventeenth century that the present 8×8 chessboard became completely established.

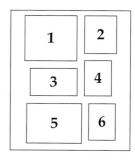

1. *Arabs playing chess*
3. *royal chess party*
5. *Moorish chess in Spain*

2. *Seven-year-old playing Go*
4. *Go formations*
6. *Caxton's 'Game of chesse'*

The Age of Exploration

The Renaissance coincided with the great sea voyages and explorations of Columbus, Vasco da Gama, Vespucci and Magellan. Such explorations necessitated the development of accurate maps and globes and of reliable navigational instruments for use at sea. These developments are described in the next few pages.

Prince Henry the Navigator (1394–1460) devoted his wealth and energies to maritime exploration. Year after year he sent ships down the west coast of Africa from his palace-observatory at Sagres in Portugal. He claimed the island groups of Madeira and the Azores and his dream was to reach a legendary Christian kingdom believed to lie in the heart of Africa.

In 1488 Bartholomew Diaz rounded the Cape of Good Hope, giving promise of a sea route to India. Ten years later **Vasco da Gama** (1469–1524) became the first European to sail around the tip of Africa and reach the west coast of India. There he made contact with Christian communities and took advantage of the lucrative spice trade.

Whereas the aims of the Portuguese explorers had been to sail south and east, their rivals the Spanish headed west, hoping to reach India by circumnavigating the globe. From 1492 **Christopher Columbus** (1451–1506), a navigator of genius, led four royal Spanish expeditions to pioneer a western route to the Indies. His expeditions reached, not India, but the new lands of North and Central America, the West Indies, and the coast of Venezuela. Further expeditions followed across the Atlantic Ocean; 'America' was named after Amerigo Vespucci, a Florentine who reached Brazil in 1502.

Ferdinand Magellan (1480–1521) led the first circumnavigation of the world, a three-year voyage that headed westward to South America, around Cape Horn, on via the Philippines (where he died in a tribal skirmish) and back via the Indian Ocean and southern Africa. Sir Francis Drake also completed a circumnavigation of the world; his astrolabe appears on page 43.

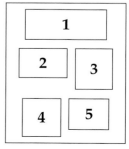

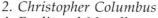

1. *Prince Henry the Navigator* 2. *Christopher Columbus*
3. *Vasco da Gama* 4. *Ferdinand Magellan*
5. *mariner with astrolabe*

Map-Making

The nautical explorations described on page 36 played a major role in the renaissance of map-making. The problem of accurately representing the spherical earth on a flat sheet of paper led to new types of map projection and to improved maps for navigators at sea. The Greek astronomer Ptolemy (see page 12), the Portuguese cosmographer Pedro Nunes, the Flemish cartographer Gerard Mercator and his friend Abraham Ortelius all played a role in this story.

Around 1500 European navigators rediscovered Ptolemy's *Geographia* and his maps came to be used extensively by navigators. Ptolemy's writings contained a detailed discussion of projections for map-making and included a 'world map' featuring Europe, Africa and Asia as well as many detailed regional maps. With the invention of printing, woodcut copies could easily be mass-produced and a number of editions appeared in the sixteenth century, each one revised to take account of new explorations.

One of the first European mathematicians to apply mathematical techniques to cartography was **Pedro Nunes** (1502–1594), Royal cosmographer and the leading figure in Portuguese nautical science. He constructed a 'nonius', an instrument for measuring fractions of a degree, and in his 1537 treatise on the sphere he showed how to represent as a straight line each rhumb line, the path of a ship on a fixed bearing.

The first world map to include America appeared in 1507, and later editions of Ptolemy included America. The first 'modern' maps of the world were due to **Gerard Mercator** (1512–1594); they can be obtained by projecting the sphere outwards on to a vertical cylinder and then stretching the map in the vertical direction. In this way, the lines of latitude (horizontal) and longitude (vertical) appear as straight lines, and all of the angles (compass directions) are correct.

In 1570 the illustrious Belgian cartographer **Abraham Ortelius** (1527–1598), geographer to the king of Spain, produced his *Theatrum orbis terrarum*, a collection of seventy maps; this is usually considered to be the first atlas. The word 'atlas' was coined by Mercator for his three-volume collection of maps in 1585–1595.

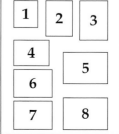

1. *Abraham Ortelius*   2. *Gerard Mercator*
3. *Gerard Mercator*   4. *Mercator projection*
5. *Pedro Nunes*   6. *Ptolemy edition, 1540*
7. *Mercator map*   8. *Ortelius map*

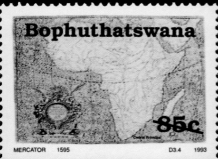

Globes

Terrestrial and celestial globes are used for representing the positions of geographical and astronomical features. During the sixteenth century, with the new interest in exploration and navigation, terrestrial globes became increasingly in demand. The first known terrestrial globe was constructed by the Nuremberg map-maker Martin Behaim in 1492. On the opposite page we present four globes from the National exhibition of mathematics and physics in Dresden.

The **Arabian celestial globe** of 1279 is one of the five oldest Islamic globes known. Constructed by Muhammed ben Mu'aijad al-Ardi of Meragha in Persia, it consists of brass overlaid with gold and silver. Only 30.5 cm high with a diameter of 14.4 cm, it illustrates the positions of about one thousand stars arranged into 47 constellations, following ideas of Ptolemy.

The **terrestrial globe** of 1568 was constructed by Johannes Praetorius of Nuremberg. It is made of brass to a scale of about 1 to 45 million, and is 47 cm high with a diameter of 28 cm. The map inscribed on the globe depicts the continents of Europe, Africa, Asia and America, with America shown joined to Asia.

The **globe clock** of 1586 was designed by Johannes Reinhold and Georg Roll of Augsburg. It is made of brass and copper covered with gold leaf, and is 56.5 cm high with a diameter of 20.5 cm. It contains a small terrestrial sphere below a large celestial sphere, all crowned by a small armillary sphere and surrounded by a movable calendar ring. The celestial sphere depicts 49 constellations.

The **heraldic celestial globe** of 1690 was designed by Erhard Weigel, a mathematics professor from Jena. It is 59 cm high with a diameter of 27.5 cm, and is made of brass covered with copper foil. It contains a system of rings representing the equator, a meridian, the tropics of Cancer and Capricorn, and other circles. Replacing the usual Ptolemaic constellations are the coat-of-arms of Saxony and other heraldic images.

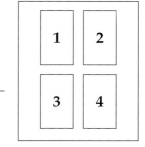

1. Arabian celestial globe 2. terrestrial globe
3. globe clock 4. heraldic celestial globe

Navigational Instruments

We have already encountered some navigational instruments—a Chinese armillary sphere (page 19), an Islamic planisphere (page 27) and an astrolabe (page 29).

Armillary spheres were usually made of metal circles representing the main circles of the universe, and were used to measure celestial coordinates or for instructional purposes. The other instruments shown here were used by navigators to measure the altitude of a heavenly body, such as the sun or the pole star, so as to determine latitude at sea.

The **astrolabe** can be traced back to Greek times, but reached its maturity during the Islamic period. It consists of a brass disc suspended by a ring that is fixed or held in the hand. On the front are various calculating devices used to take measurements of the heavenly bodies. The back has a circular scale on the rim and an attached rotating bar; to measure altitude, the observer looks along the bar at the object and reads the altitude from the scale. For navigational purposes a more basic and sturdy version was developed, known as the **mariner's astrolabe**.

Quadrants were in use in Europe from around the thirteenth century. As its name suggests, the quadrant has the shape of a quarter-circle (90°); its relations, the **sextant** and **octant**, similarly correspond to a sixth (60°) and an eighth (45°) of a circle. To measure an object's altitude, the observer views the object along the top edge of the instrument and the position of a movable rod on the circular rim gives the desired altitude.

The Jewish scholar Levi ben Gerson (1288–1344), a mathematician, philosopher and astronomer, invented the **Jacob's staff**, or cross-staff, for measuring the angular separation between two celestial bodies. Although widely used, it had a major disadvantage—to measure the angle between the sun and the horizon one had to look directly at the sun. The **back-staff** is a clever modification in which a navigator can use the instrument with his back to the sun.

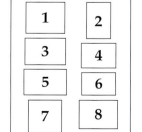

1. *armillary sphere*　　　2. *mariner's astrolabe*
3. *sextant*　　　　　　　4. *quadrant*
5. *Drake's astrolabe*　　 6. *Jacob's staff*
7. *octant*　　　　　　　8. *back-staff*

Nicolaus Copernicus

Nicolaus Copernicus (1473–1543), the 'father of modern astronomy', was born in Torun in Poland, and studied in Cracow, Bologna and Ferrara. He transformed his subject by replacing Ptolemy's earth-centred system of planetary motion (see page 12) by a heliocentric system in which the sun lies at the centre and the earth is just one of several planets travelling in circular orbits around it. Although this idea had been suggested previously by a few people, such as Aristarchus (see page 12) and Nicholas of Cusa (see page 28), Copernicus was the first to develop the underlying theory and work out its consequences in full mathematical detail.

Copernicus' book *De revolutionibus orbium coelestium* [On the revolution of the heavenly spheres] was published in 1543 and a copy of it was presented to him as he lay on his death-bed. In his book Copernicus showed that the six planets known at the time split into two groups: Mercury and Venus, whose orbits lie inside that of Earth; and Mars, Jupiter and Saturn, whose orbits lie outside it. In this way he was able to list these planets in increasing order of distance from the sun, and thereby to illuminate certain phenomena that the Ptolemaic system had failed to explain, such as why Mercury and Venus are visible only at dawn and dusk while the other planets are visible throughout the night.

The Copernican solar system aroused much controversy and brought its supporters into direct conflict with the Church who considered the earth to lie at the centre of Creation, so that Copernicus' ideas were at variance with the Holy Scriptures. The philosopher **Giordano Bruno** (1548–1600) was arrested by the Inquisition in Venice and burned alive for heresy for espousing the Copernican theory, while Galileo Galilei, at a famous Inquisition trial in 1633, was forced to recant after publishing his *Dialogue concerning the two chief world systems* which presented the Copernican system as superior to the Ptolemaic one. Galileo was not pardoned by the Church until 1995.

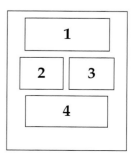

1. *Copernicus with planetary system and 'De revolutionibus'*
2. *Giordano Bruno*
3. *Copernicus portrait (Matejko)*
4. *Copernicus with title page and heliocentric diagram*

The New Astronomy

The Danish astronomer **Tycho Brahe** (1546–1601) was the greatest observer of the heavens before the invention of the telescope. In 1572 he discovered a new star, or 'nova', in the constellation of Cassiopeia, and five years later his observations of a comet proved that the solid celestial spheres on which the planets were supposed to move did not exist. In the 1570s he built the magnificent observatory of Uraniborg on the island of Hven, where he built astronomical instruments of unsurpassed accuracy and measured and catalogued over seven hundred stars.

Tycho Brahe's accurate and extensive observational records proved invaluable for his assistant **Johannes Kepler** (1571–1630). Kepler is remembered for the three laws of planetary motion from his *New astronomy* (1609) and *The harmony of the world* (1619). Tycho's data had led him to consider non-circular orbits, and Kepler proposed that:

1. the planets move in elliptical orbits with the sun at one focus;
2. the line from the sun to a planet sweeps out equal areas in equal times;
3. the square of a planet's period is proportional to the cube of its orbit's mean radius.

Kepler was indeed fascinated by ellipses and other conics, and introduced the word 'focus' into mathematics. By summing thin discs he determined the volumes of over ninety solids obtained by rotating conics and other curves around an axis, thereby foreshadowing the development of the integral calculus. Kepler was also interested in polyhedra, discovering the cuboctahedron and the antiprisms, and his name is associated with the 'Kepler-Poinsot star polyhedra'. He claimed that the five regular solids fit snugly between the orbits of the six known planets: a cube between Saturn and Jupiter, a tetrahedron between Jupiter and Mars, and so on.

Galileo Galilei (1564–1642) taught mathematics in Padua. He was the first astronomer to make extensive use of the telescope, discovering sunspots and the moons of Jupiter and drawing the moon's surface. In his mechanics book *Two new sciences* (1638) he discussed the laws of uniform and accelerated motion and explained why the path of a projectile must be a parabola.

1	2
3	4	5
6	7

1. *Kepler and planetary system*   2. *Tycho Brahe at Uraniborg*
3. *Galileo teaching at Padua*   4. *Tycho Brahe*
5. *Kepler's first two laws*   6. *Galileo's drawing of the moon*
7. *Galileo Galilei*

Calendars

Before the time of the Romans many different calendars were in use. As early as 4000 BC the Egyptians used a 365-day solar-based calendar of twelve 30-day months and five extra days added by the god Thoth. The Greek, Chinese and Jewish lunar-based calendars consisted of 354 days with extra days added at intervals, while the early Roman year had just 304 days. In 700 BC this was extended to 355 days, with the addition of the two new months Januarius and Februarius.

In 45 BC **Julius Caesar** introduced his 'Julian calendar'. This had 365¼ days, the fraction being taken care of by adding an extra 'leap day' every four years. The beginning of the year was moved to January and the lengths of the months alternated between 30 and 31 days (apart from a 29-day February in non-leap years); this was later changed by Augustus Caesar who stole a day from February to add to August and altered September to December accordingly.

Later writers determined the length of the solar year with increasing accuracy. In particular, the Islamic scholars Omar Khayyam and Ulugh Beg independently measured it as about 365 days, 5 hours and 49 minutes—just a few seconds out.

The Julian year was thus 11 minutes too long, and by 1582 the calendar had drifted by ten days with respect to the seasons. In that year **Pope Gregory XIII** issued an Edict of Reform, removing the extra days. He corrected the over-length year by omitting three leap days every 400 years, so that 2000 was a leap year, but 1700, 1800 and 1900 were not. The **Gregorian calendar** was quickly adopted by the Catholic World and other countries eventually followed suit: Protestant Germany and Denmark in 1700, Britain and the American colonies in 1752, Russia in 1917, and China in 1949.

Meanwhile, the line from which time is measured (0° longitude) was located at the Royal Observatory in Greenwich in 1884, giving rise to an **international date line** near Tonga. In 1972, atomic time replaced earth time as the official standard, and the year was officially measured as 290,091,200,500,000,000 oscillations of atomic caesium.

1	2	3
4	5	
6	7	

1. *Gregorian calendar*
2. *Gregory XIII with edict*
3. *Julius Caesar*
4. *Greenwich meridian*
5. *Greenwich observatory*
6. *international date line*
7. *reaching the millennium*

Calculating Numbers

The invention of printing in the fifteenth century led to the standardisation of mathematical notation. The **arithmetical symbols** + and − first appeared in a 1489 arithmetic text by Johann Widmann and the equals sign was introduced in a 1557 algebra book by Robert Record. The symbols × and ÷ were introduced, respectively, by William Oughtred in 1631 and John Pell in 1668.

These improvements in notation went hand in hand with developments in calculation. The Flemish mathematician **Simon Stevin** (1548–1620) wrote a popular book *De thiende* [The tenth] that explained decimal fractions, advocated their widespread use for everyday calculation, and proposed a decimal system of weights and measures. Stevin also wrote an important treatise on statics that included the first explicit use of the triangle of forces.

Johan de Witt (1625–1672) was a talented mathematician and political leader whose concern with Holland's financial problems led to his writing a treatise on the calculation of life annuity payments, an early attempt to apply probability theory to economics. His important text *Elementa curvarum linearum* [Elements of linear curves] was one of the earliest accounts of analytic geometry. He was murdered by an angry mob for political reasons.

In 1614 the Scottish laird **John Napier** (1550–1617) introduced a form of natural logarithms as an aid to mathematical calculation, being designed to replace lengthy calculations involving multiplications and divisions by easier ones using additions and subtractions. Being somewhat awkward to use, they were soon supplanted by Henry Briggs's simpler logarithms to base 10, whose use proved an enormous boon to astronomers and navigators. Later, in the 1790s, the Slovenian **Jurij Vega** (1754–1802) published a celebrated compendium of logarithms, as well as seven-figure and ten-figure logarithm tables that ran to several hundred editions. He also calculated π to 140 decimal places.

The invention of logarithms quickly led to the development of instruments based on a logarithmic scale. Most notable among these was the **slide rule**, versions of which first appeared around 1630 and were widely used for over three hundred years until the advent of the pocket calculator.

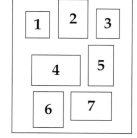

---

1. *Simon Stevin*   2. *arithmetical symbols*
3. *Johan de Witt*   4. *Napier's logarithms*
5. *slide rule*   6. *Jurij Vega*
7. *Jurij Vega*

Seventeenth-Century France

The major mathematical figures in seventeenth-century France were René Descartes, Pierre de Fermat and Blaise Pascal. The first two of these introduced algebraic methods into geometrical problems, starting a change of emphasis that would reach its climax with the work of Leonhard Euler (see page 58).

The most celebrated work of **René Descartes** (1596–1650) was his 1637 *Discours de la méthode* [Discourse on method], a philosophical treatise on universal science that contains his well-known statement 'I think, therefore I am'. The *Discours* has a 100-page appendix, *La géométrie*, containing his fundamental contributions to analytic geometry. An ancient geometrical problem of Pappus (c.300 AD) had asked for the path traced by a point moving in a specified way relative to a number of fixed lines. Descartes named two particular lengths x and y and calculated all the other lengths in terms of them, thereby obtaining a conic (parabola, ellipse or hyperbola) as the required path. Thus Descartes introduced algebraic methods into geometry, but he did not initiate the Cartesian coordinates (with axes at right angles) usually named after him. He also analysed various curves, such as the so-called **folium of Descartes** with equation $x^3 + y^3 = 3axy$, and invented a 'rule of signs' for locating the roots of a given polynomial.

Although Pierre de Fermat (1601–1665) also studied analytic geometry, he is mainly known for his contributions to number theory. These include his famous claim to have proved **Fermat's last theorem**, that for any $n > 2$ there are no non-trivial integer solutions x, y and z of the equation $x^n + y^n = z^n$. This was eventually proved by Andrew Wiles in 1995.

Blaise Pascal (1623–1662) showed an early interest in mathematics. At the age of 16, he stated his 'hexagon theorem' about six points on a conic joined in a certain way. He built a calculating machine that could add and subtract, wrote a treatise on binomial coefficients ('Pascal's triangle'), and investigated the theory of probability, atmospheric pressure ('Pascal's principle' in hydrodynamics), and the properties of such curves as the cycloid (the arch-like path traced out by a piece of mud on a bicycle tyre).

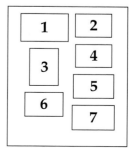

1. *Descartes and optics diagram*
2. *'Discours' (incorrect title)*
3. *Blaise Pascal*
4. *'Discours' (correct title)*
5. *folium of Descartes*
6. *Blaise Pascal*
7. *Fermat's last theorem*

Isaac Newton

The story of Isaac Newton and the apple is well known. Seeing an apple fall, he realised that the gravitational force that pulls the apple to earth is the same as the force that keeps the moon orbiting around the earth and the earth orbiting around the sun. This planetary motion is governed by a universal law of gravitation, the 'inverse-square law': the force of attraction between two objects varies as the product of their masses and inversely as the square of the distance between them; thus, if the distance is doubled, the force decreases by a factor of four. In his *Principia mathematica* (1687), possibly the greatest scientific work of all time, Newton used this law to explain Kepler's three laws of elliptical planetary motion and account for cometary orbits, the variation of tides, and the flattening of the earth at the north and south poles due to the earth's rotation.

Isaac Newton (1642–1727) was born in the tiny hamlet of Woolsthorpe in Lincolnshire in England. After attending the nearby King's School in Grantham, he went to Trinity College, Cambridge, where he avidly studied contemporary works such as a Latin edition of Descartes' *La géométrie* and a book of John Wallis on infinite series. Inspired by the latter work he produced, while still an undergraduate, the infinite series expansion for the binomial expression $(a + b)^{m/n}$—not the finite version featured on the stamp opposite, which had been known centuries earlier. Newton was later appointed Lucasian professor of mathematics at Cambridge, a post currently held by Stephen Hawking.

During the seventeenth century much progress had been made on the two branches of the infinitesimal calculus, the seemingly unrelated areas now called 'differentiation' (finding the rate at which objects move or change) and 'integration' (finding the area enclosed by a curve). It was gradually becoming realised that these processes are inverse to each other: for example, integrating a function and then differentiating the result yields our original function. During the years 1665 to 1667, Newton explored the rules of differentiation and integration and explained for the first time why the inverse relationship between these processes holds.

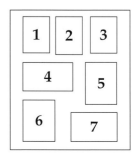

1. *apple and 'Principia' title*
2. *Isaac Newton*
3. *elliptical planetary motion*
4. *Newton's gravitation*
5. *Newton and diagram*
6. *binomial theorem*
7. *law of gravitation*

Reactions to Newton

Reactions to Newton's *Principia mathematica* were mixed. In England the book was well received, even though few readers understood it, and its reception in the Netherlands was equally positive. In France, the poet and philosopher **François-Marie Arouet de Voltaire** (1694–1778) enthusiastically supported Newton's work, and his mistress, Emilie du Châtelet, produced a French translation of the *Principia* that is still in use today.

However, not everyone in France was receptive. The *Principia* was long and difficult and raised awkward questions concerning the shape of the earth. Descartes had earlier proposed a 'vortex theory' of the universe, claiming that the planets are swept around the heavens in vortices, like corks in a whirlpool. A consequence of this theory was that the earth's rotation causes a slight elongation at the poles, so that the earth is 'lemon-shaped'. Newton had criticised the vortex theory in his *Principia*, predicting a flattening of the earth at the poles, so that the earth is 'onion-shaped'.

National pride was at stake, and settling the matter had become urgent since inaccurate map-making had led to the loss of many lives at sea. Eventually, in 1735, two geodetic missions were dispatched to settle the matter by measuring the swing of a pendulum at the equator and near the North Pole. One mission, led by **Charles Marie de la Condamine**, went to Peru and included the Spanish cosmographer **Jorge Juan** (1712–1774); the other, led by **Pierre de Maupertuis**, travelled to Lapland. These missions confirmed that Newton's theory was indeed correct: the earth is flattened at the poles.

Meanwhile, Newton's work was being criticised on another front. The Irish Bishop **George Berkeley** (1685–1753) was a vehement and highly competent critic of Newtonian science who sought to show that Newton's universe was built on quicksand. In his 1734 book *The analyst*, he unleashed a devastating attack on Newton's calculus, arguing in particular that Newton's fluxions (derivatives) were logically unsound and that his higher derivatives were the 'ghosts of departing quantities'. These difficulties were not satisfactorily resolved until the nineteenth century.

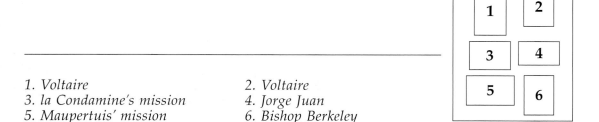

1. *Voltaire*
2. *Voltaire*
3. *la Condamine's mission*
4. *Jorge Juan*
5. *Maupertuis' mission*
6. *Bishop Berkeley*

Continental Mathematics

Although Newton could justly claim priority for the calculus, **Gottfried Wilhelm Leibniz** (1646–1716), who developed it independently, was the first to publish it. His notation, including the integral sign ∫, was more versatile than Newton's and is still used today. Leibniz's calculus was different from Newton's, being based on sums and differences rather than velocity and motion, and was part of a tradition of continental mathematics that extended from Christiaan Huygens via the Bernoulli family to Leonhard Euler.

The scientific contributions of **Christiaan Huygens** (1629–1695) were many and varied. He expounded the wave theory of light, hypothesised that Saturn has a ring, and invented the pendulum clock and spiral watch spring. In mathematics he wrote the first formal probability text, introducing the concept of 'expectation', and he analysed such curves as the cycloid and catenary (the shape formed by a hanging chain).

The Bernoulli family included several distinguished mathematicians. In his *Ars conjectandi* [Art of conjecturing], **Jakob Bernoulli** (1654–1705) proved the 'law of large numbers' that asserts that if an experiment is performed often, there is a high probability that the outcome is as expected; for example, if a fair coin is tossed two million times, the number of heads will probably be close to one million. With his brother Johann he was among the first to develop Leibniz's calculus and apply it to polar coordinates and the study of curves such as the catenary; he also invented the term 'integral'.

Leonhard Euler (1707–1783) taught in St Petersburg and Berlin and was probably the most prolific mathematician of all time. He reformulated the calculus using the idea of a function, and contributed to number theory, differential equations and the geometry of surfaces. Euler introduced the notations e (= 2.718 . . .), i (= $\sqrt{-1}$), $\sum$ (summation) and $f(x)$ (a function of x), and related the exponential and trigonometrical functions by means of the fundamental equation $e^{i\varphi} = \cos \varphi + i \sin \varphi$. He also found the 'polyhedron formula' relating the numbers of vertices, edges and faces of a polyhedron:

$$(\text{vertices}) - (\text{edges}) + (\text{faces}) = 2;$$

for the icosahedron shown opposite, $12 - 30 + 20 = 2$.

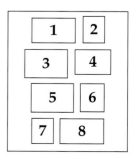

1. *Leibniz and diagram*
3. *Leibniz in Hannover*
5. *Euler in Russia*
7. *Leonhard Euler*

2. *Christiaan Huygens*
4. *Bernoulli's law of large numbers*
6. *Euler and $e^{i\varphi} = \cos \varphi + i \sin \varphi$*
8. *Euler's polyhedron formula*

Halley's Comet

Edmond Halley (1656–1742) is primarily remembered for the comet whose return he predicted and which is named after him.

While still an undergraduate at Oxford University, Halley sailed to St Helena, an island 16° south of the equator, to explore the skies of the southern hemisphere and prepare the first accurate catalogue of the stars in the southern sky. On his return he was elected to the Royal Society of London where he spent the next twenty-five years, conversing with other scientists and carrying out his researches into the solar system, comets and geophysics. In the 1680s Halley persuaded Isaac Newton to develop his ideas on gravitation and publish them in the *Principia mathematica*; Halley himself paid for the publication. In 1704 he succeeded John Wallis as Savilian professor of geometry in Oxford, where he prepared a definitive edition of Apollonius' *Conics*. Following John Flamsteed's death in 1719, Halley became Astronomer Royal and thereafter spent most of his time at the Greenwich observatory.

The earliest recorded appearances of **Halley's comet** were in 240 BC when it was reported by Chinese astronomers, and in 164 BC when Babylonian astronomers made nightly observations in their astronomical diaries. Thereafter the comet has reappeared every 75 to 80 years. It made a spectacular return in 1066, being depicted at the top of the Bayeux tapestry of the Norman conquest with the phrase *Isti mirant stella* [They wonder at the star]. The comet's appearance in 1301 inspired the Italian painter Giotto to include it in his painting *Adoration of the magi*.

Halley himself observed the comet in 1682, and realised that it was the same one that had been seen in 1531 and 1607. In 1705 he predicted its return in late 1758 or early 1759, and its appearance on Christmas Day 1758, several years after Halley's death, did much to vindicate Newton's theory of gravitation. It was on the basis of this prediction that the comet came to be named 'Halley's comet'.

1. *Edmond Halley and map*
2. *caricature of Halley as comet*
3. *Halley and Greenwich observatory*
4. *1066 comet on Bayeux tapestry*
5. *1301 comet on Giotto painting*
6. *planisphere of the southern stars*

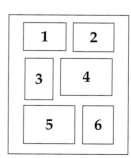

Longitude

To determine latitude north or south of the equator a mariner simply measured the angle between the sun or pole star and the horizon. However, to determine longitude east or west of home he had to compare local time with the same time at home: each hour's difference corresponds to 15° longitude, or about 1000 miles at the equator. Thus, accurate clocks were required that could be used on board a rolling ship, unaffected by changes in temperature and involving no hanging weights. Many lives were lost through inaccurate longitude readings, and in 1714 the British Parliament set up the 'Board of Longitude' which offered a £20,000 prize for a reliable timepiece.

In the absence of accurate chronometers, astronomical methods had been sought by Huygens, Newton, Halley and others. Galileo discovered a method based on eclipses of the moons of Jupiter and **Jean-Dominique Cassini** (1625–1712) obtained extensive tables of such eclipses. But it was the chronometers of **John Harrison** (1693–1776) that won the day. Over a period of forty years he constructed five timekeepers of increasing complexity. The first, Harrison's chronometer No. 1 (or H1), was used on a voyage to Lisbon in 1735, during which it erred by only a few seconds. Eventually, after much prevarication by the Board of Longitude, the prize was awarded to Harrison for his last chronometer (H5).

Meanwhile, world exploration had been continuing apace. **Louis de Bougainville** (1729–1811) made the first French circumnavigation of the world; trained in mathematics he wrote a treatise on the integral calculus. The South American plant 'bougainvillea', which he introduced to Europe, is named after him.

James Cook (1728–1779) made three major voyages. His first was a Royal Society expedition to Tahiti to observe the 1769 transit of Venus across the sun, a rare occurrence of great astronomical value; on his return trip he discovered New Zealand. On his second voyage, to Antarctica in 1772, he took Kendall's chronometer, an exact copy of Harrison's fourth timekeeper (H4), and praised it highly. Cook's third voyage ended in disaster when he was killed by the islanders of Hawaii.

1. *James Cook and sextant*
2. *Cook and transit of Venus*
3. *Jean-Dominique Cassini*
4. *Kendall's chronometer*
5. *Louis de Bougainville*
6. *Harrison's H1 chronometer*
7. *Harrison's H4 chronometer*

1	2
3	
5	4
6	7

The New World

The founders of American independence included several highly learned people who encouraged the study of mathematics and the sciences in the late eighteenth century.

Benjamin Franklin (1706–90) invented the Franklin stove, bifocal spectacles, the odometer and the lightning rod. He also carried out experiments in electricity, such as the celebrated one on lightning conduction in which he flew a kite in a thunderstorm. Although never claiming to be a mathematician, he was fascinated by magic squares and constructed a remarkable 16×16 square in which the sum of the sixteen numbers in any row, column or diagonal, or even in any 4×4 sub-square, has the value 2056.

Thomas Jefferson (1743–1826), the third president of the United States, extolled the virtues of science and wrote of the importance of calculation: extracting square and cube roots, solving quadratic equations and using logarithms. Interested in the theory and practice of classical architecture, he designed his home, Monticello, and the central rotunda of the University of Virginia. While ambassador in Paris he became interested in the metric system being proposed in France (see page 108), and strongly advocated the decimalisation of the American coinage.

Benjamin Banneker (1731–1826) was a self-taught mathematician and astronomer. When 22 years old, he designed and built an accurate striking clock, although he had never seen one previously. In later life he immersed himself in astronomy and constructed accurate astronomical tables. In 1791 he compiled the first of several almanacs, as 'the creation of a free man of the African race', and sent it to Jefferson with a plea to end slavery. Banneker was appointed by George Washington, himself a noted surveyor, to help with the surveying and layout of the new capital city.

James Garfield (1831–1881), the twentieth president of the United States, was also a keen amateur mathematician. He devised a simple but ingenious proof of Pythagoras' theorem, which later appeared in school textbooks.

Native American art has appeared on several United States stamps; one set features some geometrical designs from **Navajo blankets**.

1. *Benjamin Franklin*
2. *lightning experiment*
3. *Benjamin Banneker*
4. *Virginia rotunda*
5. *Banneker and Washington*
6. *James Garfield*
7. *Navajo blanket*
8. *Thomas Jefferson*

The Bicentennial of American Independence 1776-1976
Benjamin Franklin
11p

U.S. POSTAGE 3c
BENJAMIN FRANKLIN 250TH ANNIVERSARY

Benjamin Banneker
Black Heritage USA 15c

Jefferson 1743-1826 Virginia Rotunda
Architecture USA 15c

250th ANNIVERSARY - GEORGE WASHINGTON
E R
35c
TURKS & CAICOS ISLANDS
George Washington and Presidential Appointee Benjamin Banneker

James A. Garfield
USA 22
James A. Garfield 1881-1881

Lowe Art Museum
Navajo Art USA 22

FEDERATED STATES OF MICRONESIA
29
1743· Thomas Jefferson ·1993

France and the Enlightenment

During the seventeenth century mathematicians and scientists increasingly exchanged and communicated their ideas through publications and meetings. This led to the creation of societies and academies, such as the French **Académie des Sciences**, founded in 1666. Under the long secretaryship of Bernard Fontanelle (1657–1757), the *Académie* helped to pave the way for the scientific ideas of the Enlightenment in France.

Jean Le Rond d'Alembert (1717–1783) was a leading Enlightenment figure. He stated the ratio test for the convergence of an infinite series and attempted to formalise the idea of a limit in order to put the calculus on a firm basis. He was also the first to obtain the wave equation that describes the motion of a vibrating string. In his later years he wrote many of the mathematical and scientific articles for Denis Diderot's *Encyclopédie*, which attempted to classify the knowledge of the time.

The members of the *Académie* included Georges Louis Leclerc, **Comte de Buffon** (1707–1788) and Marie-Jean Caritat, **Marquis de Condorcet** (1743–1794). Buffon is best remembered for his 'needle experiment' which can be used to estimate π experimentally: if we throw N needles of length L randomly onto a grid of parallel lines at distance d apart, then the expected number of needles crossing a line is $2NL/\pi d$. Condorcet's main contributions were to probability theory and the study of voting patterns. After the French Revolution he was arrested while fleeing for his life and died in captivity.

Joseph Louis Lagrange (1736–1813) was Euler's successor at the court of Frederick the Great in Berlin. He wrote the first 'theory of functions', using the idea of a power series to make the calculus more rigorous, and his mechanics text *Méchanique analytique* was highly influential. In number theory he proved that every positive integer can be written as the sum of four perfect squares: for example, $79 = 49 + 25 + 4 + 1$.

Pierre-Simon Laplace (1749–1827) wrote an important text on the analytical theory of probability and is also remembered for the 'Laplace transform' of a function. His monumental five-volume work on celestial mechanics, *Traité de méchanique céleste*, earned him the title of 'the Newton of France'.

1	2
3	4
5	6	7

1. *Fontanelle and the Académie des Sciences*
2. *Jean D'Alembert*
3. *Comte de Buffon*
4. *Diderot's 'Encyclopédie'*
5. *Marquis de Condorcet*
6. *Pierre-Simon Laplace*
7. *Joseph Louis Lagrange*

The French Revolution

Out of the turbulent years of the French Revolution and the rise to power of **Napoleon Bonaparte** (1769–1821) came important developments in mathematics. Napoleon himself was an enthusiast for mathematics and its teaching, and there are even a couple of geometrical results (one involving triangles and the other involving circles) each known as 'Napoleon's theorem'.

One of Napoleon's greatest supporters was his close friend, the geometer **Gaspard Monge** (1746–1818), who accompanied him on the Egyptian expedition of 1798. Monge taught at the military school in Mézières, where he studied the properties of lines and planes in three-dimensional Cartesian geometry. While investigating possible positionings for gun emplacements in a fortress, he greatly improved on the known methods for projecting three-dimensional objects on to a plane; this subject soon became known as 'descriptive geometry'. Monge's other interests included 'differential geometry', in which calculus techniques are used to study curves drawn on surfaces, and he wrote the first important textbook on the subject.

A military genius appointed by Napoleon was the engineer **Lazare Carnot** (1755–1823), who studied under Monge at Mézières. He wrote an influential work on the 'geometry of position', in which he developed the use of vectors ('sensed magnitudes') in geometry, and was the first to assert that kinetic energy must be lost when imperfectly elastic bodies collide.

Augustin-Louis Cauchy (1789–1857) was the most important analyst of the early nineteenth century. He transformed the whole area of real analysis, providing a rigorous treatment of the calculus by formalising the concepts of limit, continuity, derivative and integral. In addition, he almost single-handedly developed the subject of complex analysis; 'Cauchy's integral formula' appears on the stamp opposite.

An important consequence of the French Revolution was the founding of the École Polytechnique in Paris. There the country's finest mathematicians, including Monge, Laplace, Lagrange and Cauchy, taught students destined to serve in both military and civilian capacities. The textbooks from the École Polytechnique were later widely used in France and the United States.

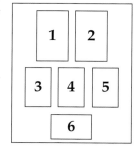

1. *Napoleon Bonaparte*   2. *Napoleon in Egypt*
3. *Gaspard Monge*   4. *Gaspard Monge*
5. *Lazare Carnot*   6. *Augustin-Louis Cauchy*

The Liberation of Geometry

Carl Friedrich Gauss (1777–1855) was one of the greatest mathematicians of all time. He lived in Göttingen and worked in many areas, ranging from complex numbers (numbers of the form $a + bi$, where $i^2 = -1$) and the factorisation of polynomials to astronomy and electricity. He investigated the construction of polygons with ruler and compass and proved that a regular n-sided polygon can be constructed when n is a 'Fermat prime number' of the form $2^{2^k} + 1$, such as 17 or 65,537. He also claimed to have discovered a 'non-Euclidean geometry', but published nothing on this.

Euclid's *Elements* (see page 10) commences with five 'postulates'. Four of these are straightforward, but the fifth postulate is different in style, resembling a theorem that ought to be provable from the others. It states that if two lines include angles x and y whose sum is less than 180°, then these lines must meet if extended indefinitely (fig. a). This result is equivalent to the 'parallel postulate', that given any line l and any point p not lying on l, there is a line parallel to l that passes through p (fig. b).

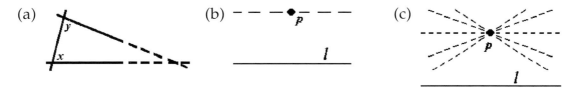

For over two thousand years mathematicians had tried to deduce this result from the first four postulates, but were unable to do so. This is because there are 'geometries' that satisfy the first four postulates but not the fifth. Such geometries have infinitely many lines parallel to l that pass through p (fig. c). They are now known as 'non-Euclidean geometries' and their existence was first published in the nineteenth century by the Russian mathematician **Nikolai Lobachevsky** (1792–1856) and the Hungarian **János Bolyai** (1802–1860). When Bolyai's father Farkas outlined János's work to his friend Gauss, the latter dismissed it as something he had already discovered; the Bolyais never forgave Gauss for this.

---

1. *Gauss and 17-sided polygon*
2. *Carl Friedrich Gauss*
3. *Gaussian (complex) number plane*
4. *Gauss and Göttingen*
5. *Farkas Bolyai*
6. *Nikolai Lobachevsky*
7. *János Bolyai*

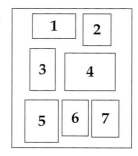

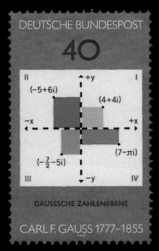

The Liberation of Algebra

Since Babylonian times, people had been able to solve quadratic equations using only arithmetic operations and the taking of roots, and in the sixteenth century Italian mathematicians developed similar solutions to cubic equations (of degree 3) and quartic equations (of degree 4). The search for a general solution to equations of degree 5 or more occupied the finest mathematicians (such as Descartes, Euler and Lagrange), but little progress was made until the Norwegian mathematician **Niels Henrik Abel** (1802–1829) proved that no such general solution can exist. A monument to Abel by the Norwegian sculptor Gustav Vigelund was erected in Oslo in 1908.

Abel's work was continued by the brilliant young French mathematician **Évariste Galois** (1811–1832), who determined which equations *can* be solved. Galois had a short and turbulent life, being sent to jail for political activism and dying tragically in a duel at the age of 20, having sat up the previous night writing out his mathematical achievements for posterity.

A breakthrough of a different nature occurred with the discovery of 'noncommutative systems', where $a \times b$ may differ from $b \times a$. **William Rowan Hamilton** (1805–1865) was a child prodigy who mastered several languages by the age of 14, discovered an error in Laplace's treatise on celestial mechanics while a teenager, and became Astronomer Royal of Ireland and Director of the Dunsink Observatory in Dublin while an undergraduate. Attempting to generalise the complex numbers, Hamilton discovered the non-commutative 'quaternions', expressions of the form

$$a + bi + cj + dk, \text{ where } i^2 = j^2 = k^2 = -1 \text{ and } ij = -ji, jk = -kj, ik = -ki.$$

Integers can be factorised as a products of prime numbers in only one way (for example, $60 = 2^2 \times 3 \times 5$), but numbers of other forms may factorise in several ways (for example, $2 \times 5 = (4 + \sqrt{6}) \times (4 - \sqrt{6})$). To circumvent this difficulty, **Richard Dedekind** (1831–1916) invented the concept of an 'ideal'; the stamp opposite shows the unique factorisation of an ideal as a product of 'prime ideals'.

The Oxford mathematician **Charles Dodgson** (1832–1898), better known as Lewis Carroll, the author of *Alice's Adventures in Wonderland*, also wrote books on the algebra of determinants, Euclidean geometry and symbolic logic.

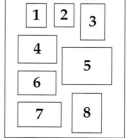

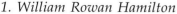

1. William Rowan Hamilton   2. Niels Henrik Abel
3. Vigelund statue   4. Hamilton's quaternions
5. Charles Dodgson   6. Évariste Galois
7. Richard Dedekind   8. Dunsink Observatory

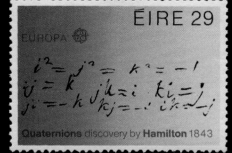

Statistics

We have already noted several contributions to probability and statistics: Johan de Witt's applications of probability to economics, Christiaan Huygens' formal probability text, Jakob Bernoulli's 'law of large numbers', Buffon's 'needle experiment' for estimating π, Condorcet's theory of voting, and the fundamental contributions of Laplace that led to his monumental *Théorie analytique des probabilités* [Analytic theory of probability].

In the early nineteenth century, the 'method of least squares' was developed by Adrien Marie Legendre and Carl Friedrich Gauss. Given a range of data arising from an experiment, the aim is to find the curve of a given type (such as a straight line or quadratic curve) that most closely fits these data; this is illustrated on the Australian stamp opposite. Gauss was led to this method while trying to predict the orbit of the newly discovered asteroid Ceres.

Adolphe Quetelet (1796–1874) was supervisor of statistics for Belgium, pioneering techniques for taking the national census. His desire to find the statistical characteristics of an 'average man' led to his compiling the chest measurements of 5732 Scottish soldiers and observing that the results were distributed normally around a mean of 40 inches. Taken with earlier studies of life annuity payments by Johan de Witt and Edmond Halley, Quetelet's investigations helped to lay the foundations of modern actuarial science.

Florence Nightingale (1820–1910) was strongly influenced by the work of Quetelet. Mainly remembered as the 'lady with the lamp' who saved many lives through her sanitary improvements in hospitals during the Crimean War, she was also a fine statistician who collected and analysed mortality data from the Crimea and displayed them using her 'polar diagrams', a forerunner of the pie chart.

A number of stamps have featured statistical themes—for example:

- an Egyptian stamp featuring Amenhotep (see page 4), issued for a 1927 statistical congress in Cairo;
- population graphs portraying the age distribution of men and women in various years;
- a graph of the Norwegian gross national product from 1876 to 1976.

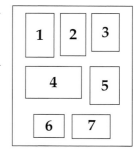

1. *Florence Nightingale*
2. *Amenhotep*
3. *Adolphe Quetelet*
4. *population graph*
5. *population graph*
6. *gross national product*
7. *best-fit curve*

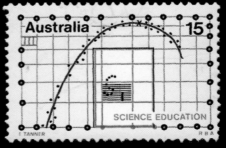

China and Japan

The first missionary in China was the Italian Jesuit **Matteo Ricci** (1552–1610), who disseminated knowledge of western science—especially in mathematics, astronomy and geography. His most important contribution was an oral Chinese translation of the first six books of Euclid's *Elements*, which was recorded by **Xu Guangqi** (1562–1633), Grand Secretary of the Wen Yuan Institute and 'the first man in China after the monarch himself'. By the time of Ricci's arrival near the end of the Ming dynasty, the calendar of Guo Shoujing (see page 16) had become very inaccurate; Xu Guangqi was appointed to direct its reform.

In 1742 Christian Goldbach wrote to Leonhard Euler conjecturing that every even integer can be written as the sum of two prime numbers (for example, $18 = 13 + 5$ and $20 = 17 + 3$). Although Goldbach's conjecture remains unresolved, a partial result of Chen Jing-Run (1966), shown on the stamp opposite, implies that every sufficiently large even integer can be written as the sum of a prime number and a number with at most two prime factors.

Number theory was also the province of **Hua Loo-Keng** [Hua Luogeng] (1910–1985), who wrote important texts on the subject. During the Cultural Revolution of 1966–1976, he travelled widely through China lecturing on industrial mathematics to audiences of up to 100,000 factory workers.

Determinants are algebraic expressions that arise in the solution of systems of simultaneous linear equations. The Japanese mathematician **Takakazu Seki** (1642–1708), also known as 'Seki Kowa', was the first mathematician to investigate determinants, a few years before Leibniz (who is usually given priority) contributed to the subject. In 1683 Seki explained how to calculate determinants up to size 5×5, and the Japanese stamp shows his diagram for calculating the products that arise in the evaluation of 4×4 determinants.

The Japanese abacus, or 'soroban', appears in a detail from the *Three beauties* by Toyokuni Utagawa, around 1800. The stamp opposite was issued for the Thirteenth World Congress of Certified Public Accountants in Tokyo in 1987.

Shogi, also called 'the game of the generals', is a Japanese form of chess that probably reached Japan from Korea in the eighth century. The board is of size 9×9, and each player has twenty pieces that are placed on the squares.

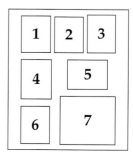

1. *Matteo Ricci*
2. *Japanese abacus*
3. *Takakazu Seki*
4. *Xu Guangqi*
5. *Hua Loo-Keng*
6. *Goldbach's conjecture*
7. *Japanese shogi*

Russia

The nineteenth and twentieth centuries produced many fine Russian mathematicians, such as Nikolai Lobachevsky, co-founder of non-Euclidean geometry (see page 70).

Mikhail Ostrogradsky (1801–1861) took up mathematics to finance his ambition of becoming an army officer. While studying in Paris, he worked in mathematical physics, proving the 'divergence theorem', often ascribed to Gauss. He later achieved his ambition by teaching mathematics in the Russian military academies.

A generation later was **Pafnuty Chebyshev** (1821–1894), remembered mainly for his work on orthogonal functions ('Chebyshev polynomials'), probability ('Chebyshev's inequality'), and prime numbers. He founded the St Petersburg School of Mathematics. **Aleksandr Lyapunov** (1857–1918) was much influenced by Chebyshev. He worked on the stability of rotating liquids and the theory of probability. In 1918 his wife died of tuberculosis; Lyapunov shot himself the same day and died shortly after.

Nikolai Zhukovsky (1847–1921) studied the flow of a fluid inside and around solid objects. Considered 'the father of Russian aviation', he analysed the flow of air around an aeroplane wing, thereby making it possible to solve the problem of lift. **Konstantin Tsiolkovsky** (1857–1935), a pioneer of rocket flight, invented multi-stage rockets and produced a celebrated law that relates the velocity and the mass of a rocket in flight.

The mathematician and novelist **Sonya Kovalevskaya** (1850–1891) made valuable contributions to mathematical analysis and partial differential equations. Barred by her gender from studying in Russia, she went to Heidelberg, attending lectures of Kirchhoff and Helmholtz, and to Berlin, later becoming the first female professor in Stockholm. She won the coveted Prix Bordin of the French Academy for a memoir on the rotation of bodies.

Otto Schmidt (1891–1956) was a distinguished algebraist and polar explorer who wrote a celebrated book on the theory of groups and was rescued from a stranded ice-breaker during a polar scientific expedition.

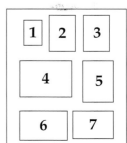

1. *Aleksandr Lyapunov*   2. *Mikhail Ostrogradsky*
3. *Pafnuty Chebyshev*   4. *Konstantin Tsiolkovsky*
5. *Sonya Kovalevskaya*   6. *Otto Schmidt and polar camp*
7. *Nikolai Zhukovsky*

Eastern Europe

Ruđer Boščović [Ruggiero Boscovich] (1771–1787) was born in Ragusa (now Dubrovnik). He was interested in astronomy, gravitation, trigonometry and optics, and was the first to compute a planet's orbit from just three observations of its position. Boščović anticipated later atomic theory by regarding atoms, not as hard indivisible entities as Democritus had assumed (see page 6), but as centres of force. He also developed an early version of the method of least squares in statistics.

Bernhard Bolzano (1781–1848) was born in Prague. Like Cauchy (see page 68), he was instrumental in formalising the idea of a continuous function (one whose graph has no 'jumps'), and proved the 'intermediate value theorem', that the graph of a continuous function takes on all values between the lowest and highest points. He also obtained a function that is continuous everywhere but differentiable nowhere (every point is a 'corner'). Living in Prague, he was isolated from the mainstream of European mathematical activity and his contributions unfortunately had little impact.

Hermann Oberth (1894–1989), shown opposite explaining an aspect of planetary theory, was born in Transylvania, now Romania. He published theories on how a rocket could function in a vacuum and explained that, given sufficient thrust, a rocket could encircle the earth.

The **Union of Czechoslovak Mathematicians and Physicists** was founded in 1862, and its 125th anniversary was commemorated by a set of three stamps. One of these features the Prague astronomical clock and a computer graphic from the theory of functions, superimposed on a result from real analysis.

The Romanian monthly *Gazeta matematica* was first published in 1895. With its aim of developing the mathematical knowledge of high-school students, it has had an enormous influence on mathematical life in Romania. A centenary stamp featuring **Ion Ionescu**, its 'Spiritus rector' and editor for over fifty years, was issued in 1995.

During the Second World War the Germans' Enigma codes were broken by a team at Bletchley Park in England, led by Alan Turing. These code-breaking methods refined earlier ones of Polish cryptographers whose contributions are commemorated on a **Polish Enigma** stamp.

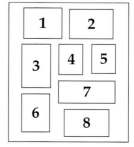

1. *Ruđer Boščović*
3. *Ion Ionescu*
5. *Bernhard Bolzano*
7. *Czech mathematical society*
2. *Hermann Oberth*
4. *Ruđer Boščović*
6. *Enigma code-breakers*
8. *Gazeta matematica*

Mathematical Physics

If a compass needle is placed near a wire that carries electrical current, it is deflected. This observation, relating electricity and magnetism, led to the subject of 'electromagnetism'. Two French mathematical physicists who made early contributions in this area were **André-Marie Ampère** (1775–1836) and **François Arago** (1786–1833). Ampère is best remembered for his work in electricity: the SI unit of current flow is named after him. He was the first to apply advanced mathematics to magnetic and electrical phenomena, and 'Ampère's law' relates the magnetic field between two wires to the product of the currents in them. Arago taught analytic geometry at the École Polytechnique (see page 68), and worked on mechanical linkages, magnetism and the polarisation of light.

In 1850 another French physicist, **Jean Foucault** (1819–1868) carried out his famous **pendulum experiment**, designed to show the rotation of the earth on its axis. A 28kg ball was suspended from the dome of the Panthéon in Paris and allowed to swing. As the earth rotated, the swinging pendulum's path gradually shifted, finally returning to its starting point 24 hours later. Foucault also measured the velocity of light using a rapidly rotating mirror.

In Germany, the astronomer and mathematician **Friedrich Wilhelm Bessel** (1784–1846) made measurements on over 50,000 stars and was the first to determine interstellar distances accurately by using the method of parallax. In 1817, while investigating a problem of Kepler, he introduced the Bessel functions $J_n(x)$ which satisfy a certain second-order differential equation. The German stamp opposite features the graphs of the Bessel functions $J_0(x)$ and $J_1(x)$. Bessel functions have wide application throughout physics—notably in acoustics (the vibration of a drum) and electromagnetism.

Bessel functions also arise in the solution of Helmholtz's wave equation when cylindrical or spherical symmetry is involved. **Hermann von Helmholtz** (1821–1894) was a German/Prussian physicist, mathematician and biologist who invented the ophthalmoscope for viewing the eye's retina, formulated the general law of the conservation of energy, and wrote extensively on the mathematics of sound and electromagnetic waves.

1. *Friedrich Bessel*
3. *Jean Foucault*
5. *André-Marie Ampère*
7. *Foucault's pendulum*

2. *Friedrich Bessel*
4. *François Arago*
6. *Hermann von Helmholtz*

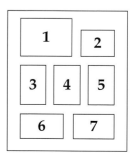

The Nature of Light

In 1831 Michael Faraday carried out his famous experiments that led to the discovery of 'electromagnetic induction', where electricity is induced by changes in a magnetic field.

Using the most advanced vector methods of his day, **James Clerk Maxwell** (1831–1879) synthesised Faraday's laws of electromagnetism into a coherent mathematical theory, confirming Faraday's intuition that light consists of electromagnetic waves. His celebrated *Treatise on electricity and magnetism*, containing the fundamental mathematical laws of electromagnetism now known as 'Maxwell's equations', predicted the existence of such phenomena as radio waves. **Heinrich Hertz** (1857–1894) confirmed this: his 'Hertzian waves' later formed the basis of Marconi's work on radio telegraphy.

Hendrik Lorentz (1853–1928) showed how Maxwell's electromagnetic waves interact with matter consisting of atoms within which are distributions of electric charge. He predicted that magnetic fields modify the spectral lines of atoms; this was confirmed by his pupil Pieter Zeeman, with whom he shared the 1902 Nobel Prize for Physics.

Maxwell's electromagnetic theory of light, and experiments on the newly invented light bulb, led scientists to consider how atoms emit light. At first it seemed that all the light should have very high frequency. Reconciling theory with experiment, **Max Planck** (1858–1947) announced the first steps towards 'quantum theory' by postulating that atoms can emit light only in small packets (called 'quanta') whose energy E is proportional to their frequency ν—thus, $E = h\nu$, where h is 'Planck's constant'. Lacking sufficient energy to make many high-frequency packets, atoms would be forced to radiate at lower frequencies instead.

In 1905 **Albert Einstein** (1879–1955) explained the 'photoelectric effect', that light behaves like particles and can liberate electrons on impact with a metal surface. Einstein's paper, which led to his 1921 Nobel Prize in Physics, showed that Planck's equation $E = h\nu$ is a fundamental feature of light itself, rather than of the atoms. Despite the many experiments that show light to consist of waves, it also behaves as though it consists of particles, each carrying one quantum ($h\nu$) of energy.

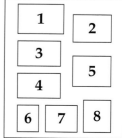

1. *Maxwell's equations*
3. *Heinrich Hertz*
5. *Hertzian radio waves*
7. *Max Planck*

2. *James Clerk Maxwell*
4. *photoelectric effect*
6. *Hendrik Lorentz*
8. *Planck's constant*

Einstein's Theory of Relativity

Since the eighteenth century it had been assumed that all mechanical motion throughout the universe, whether of planets, apples or atoms, satisfies the laws of motion expounded by Isaac Newton in his *Principia mathematica* (see page 54). Moreover, these laws still hold for systems moving at a constant speed relative to each other: thus, if a smoothly moving train travels at constant speed, the passengers cannot tell whether they are in motion unless they look out of the window; but if the train accelerates, then the motion is noticed.

In 1905 **Albert Einstein** published his 'special theory of relativity'. Until that time it had been assumed that Maxwell's equations (see page 84) are valid only in a particular frame of reference (that of the 'ether' that carries the waves) and are thus unlike Newton's laws which hold for all observers in uniform motion. Einstein reconciled this apparent discrepancy by starting from the simple postulate that the basic laws of physics (including Maxwell's equations) are the same for all observers in uniform motion relative to one another.

Thus, Einstein was able to extend Newton's ideas on mechanics so as to incorporate electromagnetism and the results of Maxwell. The equations that Newton had used to get from one frame of reference to another were now replaced by new ones due to Lorentz. A consequence of these ideas is that mass is simply a form of energy, and that the energy E and the mass m are related by the well-known relativity equation $E = mc^2$, where c is the speed of light.

Ten years later, in his 'general theory of relativity', Einstein extended his earlier ideas to include accelerated motion and gravity. His starting point was the observation that for passengers in a uniformly accelerated system, such as a spaceship, the perceived force is indistinguishable from gravity. Einstein's general theory used some newly developed ideas from higher-dimensional differential geometry, and in turn stimulated further developments in that subject.

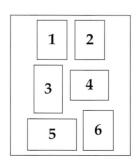

1. Albert Einstein
3. Einstein with violin
5. Einstein's law
2. Albert Einstein
4. Einstein quotation
6. Einstein with equation

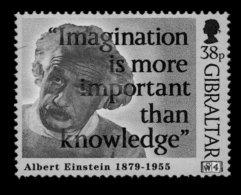

Quantum Theory

Shortly after Einstein's paper on the photoelectric effect, Ernest Rutherford's experiments led to the idea of an atom as a 'miniature solar system' in which the electrons orbit around a nucleus. To explain why atoms radiate light at the particular frequencies observed, **Niels Bohr** (1885–1962) proposed that the angular momentum of each orbiting electron must be an exact multiple of the basic unit $h/2\pi$.

In 1923 **Louis de Broglie** (1892–1987) suggested that if light waves can behave as particles, then perhaps other particles (such as electrons) can also behave as waves; this was rapidly verified experimentally. In de Broglie's theory Bohr's electron orbits are precisely those whose circumference is an integer number of wavelengths.

Twenty-five years after Planck's original quantum hypothesis, physicists still had no coherent understanding of quantum theory, but in 1925 two independent discoveries revolutionised the subject. **Erwin Schrödinger** (1887–1961) learned of de Broglie's idea that particles can also behave like waves, and found the appropriate partial differential equation to describe these waves. Earlier, **Werner Heisenberg** (1901–1976) had taken an algebraic approach, realising that quantum-theoretical quantities such as position, momentum and energy can be represented by infinite arrays, or 'matrices'. Using the fact that matrix multiplication is non-commutative, Heisenberg deduced his famous 'Uncertainty Principle', that it is theoretically impossible to determine the position and the momentum of an electron at the same time.

In 1928 **Paul Dirac** (1902–1984) effectively completed classical quantum theory by deriving an equation for the electron that, unlike those of Schrödinger and Heisenberg, was consistent with Einstein's theory of relativity. This equation explained electron spin, and led him to predict the existence of antiparticles, such as the positron which was detected four years later. Schrödinger discovered that, despite the apparent differences, his theory was equivalent to that of Heisenberg.

In the period 1929 to 1933 Nobel Prizes in Physics were awarded to de Broglie, Heisenberg, Dirac and Schrödinger for their contributions to quantum mechanics.

1. *Niels and Margrethe Bohr*   2. *Louis de Broglie*
3. *Werner Heisenberg*   4. *Paul Dirac*
5. *Paul Dirac*   6. *de Broglie's law*
7. *Erwin Schrödinger*

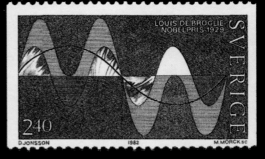

The Twentieth Century

King Oscar II of Sweden and Norway (1829–1907) was an enthusiastic patron of mathematics. On his sixtieth birthday he offered a prize of 2500 Swedish crowns for a memoir on mathematical analysis. The winner was **Henri Poincaré** (1854–1912), who wrote on the 'three-body problem' of determining the motion of the sun, earth and moon. Arguably the most brilliant mathematician of his generation, and a gifted populariser of the subject, Poincaré contributed to many areas of mathematics and physics, including celestial mechanics, differential equations and algebraic topology.

Bertrand Russell (1872–1970) was one of the outstanding figures of the twentieth century. One version of his celebrated 'Russell paradox' of 1902 asks: 'in a village the barber shaves all those who don't shave themselves; who shaves the barber?' In 1913 he and A. N. Whitehead completed *Principia mathematica*, a pioneering work on the logical foundations of mathematics.

Srinivasa Ramanujan (1887–1920) was one of the most intuitive mathematicians of all time. Mainly self-taught, he left India in 1914 to work in Cambridge with G. H. Hardy, producing some spectacular joint papers in analysis and number theory before his untimely death at the age of 32. While visiting him in hospital Hardy remarked that his taxi number 1729 was very uninteresting. 'No', replied Ramanujan, 'it's the smallest number we can write as the sum of two cubes in two ways!' [1000 + 729 and 1728 + 1]

Three analysts featured on stamps are **Constantin Carathéodory** (1873–1950), **Stefan Banach** (1892–1945) and **Norbert Wiener** (1894-1964). Carathéodory worked on the theory of functions and the calculus of variations, while Banach helped to create modern functional analysis and develop links between topology and algebra; the term 'Banach space' is named after him. Wiener applied his analytical results to statistics and potential theory and invented 'cybernetics', the study of communication processes.

When a recurrence of the type $z_{n+1} = z_n^2 + c$ is applied to each point z_0 in the complex plane, the boundary curve between those points that remain finite and those that 'go to infinity' is a fractal pattern, called a **Julia set** after the French mathematician Gaston Julia. The stamp opposite shows a detail of the Julia set that arises when $c = 0.2860 + 0.0115i$.

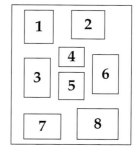

1. *Bertrand Russell*	2. *Srinivasa Ramanujan*
3. *Julia set fractal*	4. *Norbert Wiener*
5. *King Oscar II*	6. *Henri Poincaré*
7. *Constantin Carathéodory*	8. *Stefan Banach*

The Birth of Computing

The first mechanical calculating machines appeared in the seventeenth century, following the invention of logarithms by John Napier and the development of the slide rule (see page 50). An early machine was described by **William Schickard** in 1623, and later calculating machines were constructed by Pascal and Leibniz. Pascal's machine was operated by cogged wheels and could add and subtract, while Leibniz' machine could also multiply and divide; neither machine worked efficiently in practice.

The central figure of nineteenth-century computing was **Charles Babbage** (1791–1871), who may be said to have pioneered the modern computer age with his 'difference engine' and 'analytical engine'. The difference engine, a digital machine conceived in the 1820s, was a highly complex arrangement of gears and levers designed to mechanise the calculation of mathematical tables and print out the results. Although a small working model was built, a full-scale version was not constructed from Babbage's detailed drawings until 1991.

Babbage's analytical engine can be regarded as the forerunner of the modern programmable computer, even though it was never actually constructed. Designed to be run by steam power, it contained a store (or memory) and was to be programmed by means of punched cards.

The idea of a **punched card**, with holes punched in specific locations to convey information, had been used in the 'Jacquard loom' of **Joseph Marie Jacquard** (1752–1834) in order to mechanise the weaving of complicated patterns. Data processing with punched cards was developed by Herman Hollerith for the US population census of 1890, and along with **punched tape**, such cards were in widespread use for many years afterwards.

The modern computer age started in the Second World War, with COLOSSUS in England, used for deciphering German military codes, and **ENIAC** [Electronic Numerical Integrator And Computer] in the United States. These machines were large and cumbersome: ENIAC was eight feet high and contained 17,000 vacuum tubes, 70,000 resistors, 10,000 capacitors, 1500 relays and 6000 switches. Inspired by ENIAC, **John von Neumann** (1903–1957) contributed to the development of a 'stored program computer' in which the data and instructions are held in an internal store until needed.

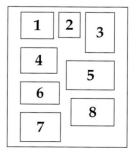

1. *Schickard's machine*
2. *Joseph Marie Jacquard*
3. *Charles Babbage*
4. *punched tape*
5. *ENIAC computer*
6. *punched card*
7. *computer operator*
8. *John von Neumann*

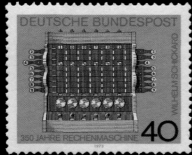

The Development of Computing

Since the 1950s computers have developed at an ever-accelerating pace, with a massive increase in speed and power and a corresponding decrease in size and cost. They have become so much a part of our everyday life that it is difficult to remember what life was like beforehand.

The 'first generation' of electronic digital computers spanned the 1950s. These computers stored their programs internally and initially used vacuum tubes as their switching technology. However, such tubes were bulky, hot and unreliable, and were gradually replaced in the 'second generation' of computers by transistors. This development in turn gave rise to the problem of trying to interconnect many thousands of simple circuits to form a system with sufficient computing power.

The 'third generation' of computers, in the late 1960s, saw the solution of this problem with the development of printed circuit boards on which thin strips of copper were 'printed', connecting the transistors and other electronic components. This led to the all-important introduction of the **integrated circuit**, an assembly of thousands of transistors, resistors, capacitors and other devices, all interconnected electronically and packaged as a single functional item. In the 1970s the first personal computers became available, for use in the home and office.

The invention of the **World Wide Web** by **Tim Berners-Lee** in the early 1990s has led to the information superhighway, whereby all types of information from around the world are easily accessible. Communications have also been transformed with the introduction of electronic mail; the Thailand stamp portrays **King Bhumibol** checking his e-mail.

Computer-aided design has also developed rapidly, and in 1970 the Netherlands produced the first set of computer-generated stamp designs; the stamp opposite shows an **isometric projection** in which the circles at the centres of the faces gradually expand and become transformed into squares. The **computer drawing** of a head is a graphic from the 1972 computer-animated film *Dilemma*.

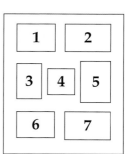

1. *visual display unit*
2. *King Bhumibol at e-mail*
3. *isometric projection*
4. *World Wide Web*
5. *computer drawing*
6. *integrated circuit*
7. *Tim Berners-Lee*

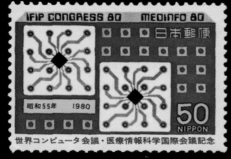

The International Scene

A s part of the four-hundredth anniversary celebrations of Columbus's 1492 voyage to America, a World Congress of Mathematicians took place at the World's Columbian Exposition in **Chicago** in 1893. Only forty-five mathematicians attended and the opening address, on 'The present state of mathematics', was given by Felix Klein of Göttingen, one of just four participants from outside the US. Since that first meeting, more than twenty international congresses have been held, usually every four years.

In recent times, several of the International Congresses of Mathematics have been commemorated by stamps. The first of these was the **Moscow** congress of 1966, attended by 4000 mathematicians from forty-nine countries. Twelve years later, the congress in **Helsinki** attracted delegates from eighty-three countries and the stamp design featured differential geometry.

In 1982 a set of four stamps was issued to commemorate the International Congress in **Warsaw**. This set featured the Polish mathematicians Stefan Banach (see page 91), Waclaw Sierpinski, Stanislaw Zaremba and Zygmunt Janiszewski. In the event, the uncertain political situation necessitated the postponement of the congress to 1983.

The first International Congress to be held outside Europe or North America took place in **Kyoto** in 1990. The stamp design shows an origami polyhedron. In 1994 the congress was held for the third time in Zürich. The commemorative stamp featured Jakob Bernoulli and his law of large numbers (see page 57). For the 1998 congress in **Berlin** the stamp design included a solution of the 'squaring-the-rectangle' problem of dividing a rectangle with integer sides (in this case, 176 and 177) into unequal squares with integer sides. The background consisted of spirals made from the decimal digits of π.

Other international mathematical events have been commemorated on stamps, including the Second European Congress of Mathematics, held in **Budapest** in 1996. The International Union of Mathematics, supported by UNICEF, declared the year 2000 as **World Mathematical Year**, and a number of countries issued special stamps.

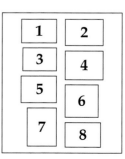

1. *Chicago 1893*
3. *Helsinki 1978*
5. *Kyoto 1990*
7. *Budapest 1996*

2. *Moscow 1966*
4. *Warsaw 1982/3*
6. *Berlin 1998*
8. *World Mathematical Year 2000*

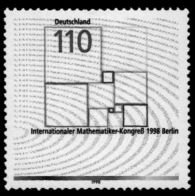

Mathematics and Nature

M athematics occurs throughout nature—from the arithmetic of sunflowers and ammonites to the geometry of crystals and snowflakes.

The Fibonacci sequence 1, 1, 2, 3, 5, 8, 13, . . . (see page 28) occurs throughout nature. If we count the leaves around a plant stem until we find one directly above our initial leaf, the number of intervening leaves is generally a Fibonacci number (5 for oak trees, 8 for poplar trees, 13 for willow trees, etc.). The spiral arrangements of scales on a **pine cone** similarly exhibit 8 right-hand and 13 left-hand spirals, while much larger Fibonacci numbers (34, 55, etc.) appear in the spiral arrangements of seeds in a **sunflower** head.

The ratios of successive terms $\frac{1}{1}, \frac{2}{1}, \frac{3}{2}, \frac{5}{3}, \ldots$ of the Fibonacci sequence tend to a limit, often called the 'golden ratio' and equal to $\frac{1}{2}(1 + \sqrt{5}) = 1.618.\ldots$ This ratio arises throughout mathematics: it is the ratio of a diagonal and a side of a regular pentagon, and a rectangle with sides in this ratio is often considered to have the most pleasing shape. Moreover, the removal of a square from one end of a golden rectangle leaves another such rectangle; this process is shown on the Swiss stamp opposite, which also features the closely related **logarithmic spiral**, found on snail shells and **ammonites**.

The delicate structure of a **snowflake** has six-fold rotational symmetry—rotation by 60° leaves the pattern unchanged—and no two snowflakes are exactly the same. The hexagonal form of snowflakes was recognised by the Chinese in the second century BC and was later investigated by Johannes Kepler, René Descartes and others.

There are three regular types of tiling pattern (or tessellation) for the plane: those formed by equilateral triangles, squares and regular hexagons. The hexagonal tiling pattern appears in nature in the form of a bees' **honeycomb**; a hexagonal stamp featuring bees on a honeycomb appears on page 111.

As liquids crystallise they assume the form of a polyhedron. There are many types of crystal shape: **fluorite crystals** appear as regular octahedra (eight triangular faces), while **lead sulphide crystals** appear as cuboctahedra (eight triangular faces and six square faces) and **zinc sulphide crystals** appear as truncated tetrahedra (four triangular faces and four hexagonal faces).

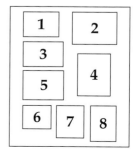

1. *logarithmic spiral*   2. *snowflake*
3. *pine cone*   4. *ammonite*
5. *sunflower*   6. *fluorite crystals*
7. *sulphide crystals*   8. *honeycomb*

Twentieth-Century Painting

In the late nineteenth and early twentieth century, artists, musicians, poets and novelists were fascinated with the fourth dimension and how to portray it. Their interest in mathematical ideas, such as non-Euclidean geometry, as well as in recent scientific discoveries such as electromagnetic waves, X-rays and electrons, provoked artists to look at the world in a new way and record their observations in paintings and sculpture.

Robert Delaunay (1885–1941) rejoiced in the visual impact of colours. In his 1931 painting *Rhythme: Joie de vivre* the vivid circles represent the haloes around glowing electric street lights. Interpreting this work, Delaunay remarked 'Tout est halo' [All is halo], adding that he had never seen a straight line in his life. Delaunay and his contemporaries were aware of the new geometries and of the writings of Henri Poincaré (see page 90).

Piet Mondrian (1872–1944) tried to portray the universal mathematical harmonies behind visual phenomena. In his 1943 painting *Broadway Boogie-Woogie*, the jazz-loving Mondrian created spatial ambiguity by means of coloured lines in a rectangular arrangement. For years he had shared the enthusiasms of other artists in trying to portray the fourth dimension, and in his earlier paintings of white and coloured rectangles separated by straight black lines, the white space represents the fourth dimension. Eventually he rejected three-dimensional visual reality as a superficial illusion, believing that two dimensions represent objects clearly whereas a perspective representation (implying three dimensions) weakens them. Consequently, he ensured that all overlapping shapes and all diagonals (which could be interpreted as three-dimensional) were eliminated from his work.

Optical art stimulates and confuses our visual perception. **Victor Vasarely** (1908–1997) was much influenced by the work of Mondrian, and studied colour theory, perception and illusion. In *Vega-chess*, he made use of this illusion by repeating the same motif over and over again, making it difficult for us to see the picture. The painting, which seems to have no background, can be hung on the wall with any side at the top, because of its 4-fold rotational symmetry. His 1975 design *Tetcye* is similarly based on symmetry: in this case, 6-fold rotational symmetry.

1	2
3	4

1. Vasarely's 'Tetcye'
2. Delaunay's 'Rhythme: Joie de vivre'
3. Vasarely's 'Vega-Chess'
4. Mondrian's 'Broadway Boogie-Woogie'

The Geometry of Space

The geometry of space takes many forms: in the design of buildings, in the symmetry of bridges, or in the sculptures that decorate our towns and cities.

The Dutch graphic artist **Maurits Escher** (1898–1972), the 'master of optical illusion', has had an enormous influence on geometrical art. He is best known for his intricate tiling patterns of birds and beasts, angels and devils, or fish and boats. A celebrated example is the 42-foot mural *Metamorphose* in the Kerkplein post office in The Hague; the stamp opposite shows Escher with part of this mural, depicting a tiling arrangement of reptiles. Escher also designed several postage stamps, including the **posthorn design** used for the seventy-fifth anniversary of the Universal Postal Union in 1949.

Escher's three-dimensional designs exhibit a mastery of perspective. Several incorporate 'impossible' features that deliberately distort the scene, showing staircases that forever ascend or descend, or objects that cannot exist in space. The Austrian stamp opposite depicts an **impossible cube** inspired by his designs, while the Swedish stamps show an **impossible triangle** and an **impossible object** drawn in 1934 by the Swedish artist Oscar Reutersvärd.

Another interesting geometrical object is the **Möbius strip**, named after the German mathematician and astronomer August Möbius in 1858. It is shown opposite and has only one side and one boundary edge. To construct it, take a long rectangular strip of paper, twist one end of it through 180° and then glue the ends together.

An attractive three-dimensional sculpture in the form of a Möbius strip is *Continuity* by the Swiss architect Max Bill. It can be seen in front of the Deutsche Bank in Frankfurt, and was carved from a single piece of granite weighing 80 tonnes. Another sculpture, this time in the form of an enormous helix, is the Brazilian sculpture *Expansion*, symbolising progress.

A 'ruled surface' is a curved surface constructed from closely packed straight lines, sometimes in the form of a hyperbolic paraboloid. Several buildings use such surfaces for their roofs; an example is the **German pavilion** constructed for the 1997 world fair in Montréal.

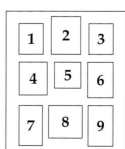

1. *impossible triangle*
2. *Escher's impossible cube*
3. *impossible object*
4. *sculpture 'Continuity'*
5. *Escher posthorn design*
6. *Escher with mural*
7. *sculpture 'Expansion'*
8. *German pavilion*
9. *Möbius strip*

Mathematical Games

The position game of **noughts-and-crosses**, or 'tic-tac-toe', developed in the nineteenth century from earlier 'three-in-a-row' games such as three men's morris. Two players take turns to place their symbol (O or ×) in a square and try to get three in a row. Variations involve larger boards (4 × 4 or 5 × 5) and three or more dimensions (4 × 4 × 4 or 3 × 3 × 3 × 3).

Other board games in which mathematical strategy plays a part are **fox-and-geese** and **Chinese checkers**. Fox-and-geese, a 'hunt game' with horizontal, vertical and diagonal moves, probably originated around the fourteenth century. One version has thirteen 'geese' who try to catch a 'fox'; the fox can jump over the geese and remove them, while the geese try to surround the fox. Chinese checkers is played by two to six players on a star-shaped board. The object is to move one's pieces from one point of the star to the opposite point while trying to prevent one's opponents from doing the same.

Until 1000 AD, most labyrinths were of the classic seven-ring form (see page 15). In the ensuing centuries, they gave rise to increasingly complicated **mazes**, which are now to be found everywhere from puzzle books to ornamental gardens. In the late nineteenth century algorithms for escaping from a maze were presented by Gaston Tarry and others.

The game of **dominoes** first came to be widely played in the eighteenth century. Around 1850 several mathematicians sought the number of different ways of placing the 28 dominoes (0–0 to 6–6) in a ring, and in 1871 M. Reiss obtained this number; it is over three million.

The **Rubik cube**, invented in 1974 by the Hungarian engineer Erno Rubik, is a 3 × 3 × 3 coloured cube whose six faces can be independently rotated so as to yield 43,252,003,274,489,856,000 different patterns. Given such a pattern, the object is to restore the original colour of each face. In the early 1980s, when the Rubik cube craze was at its height, over 100 million cubes were sold and public cube-solving contests were held in Hungary and elsewhere.

In the past twenty years computers have been developed that play chess and other games. At first these were rudimentary, but now they regularly beat the chess masters. **Bridge** is also a game of strategy; it is a card game played by four players, and is strongly based on the theory of probability.

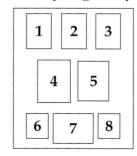

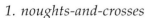

1. *noughts-and-crosses*   2. *maze*
3. *dominoes*   4. *Rubik cube*
5. *computer chess*   6. *fox-and-geese*
7. *bridge*   8. *Chinese checkers*

Mathematics Education

We have already noted several varied examples of mathematics teaching and education, at both elementary and advanced levels: the teaching of scribes and accountants in ancient Egypt, the role of the mathematical arts at Plato's Academy, Euclid and his pupils, the role of the 'quadrivium' as part of the curriculum for the first European universities, the appearance of the first printed mathematical textbooks, Galileo teaching in Padua, the founding of the École Polytechnique in Paris after the French Revolution, the education of Sonya Kowalevskaya, Hermann Oberth teaching planetary theory, and the role of the *Gazeta matematica* in the teaching of mathematics in Romania.

A substantial number of stamps have appeared featuring the teaching of elementary mathematics—for example:

- a stamp from Guinea-Bissau, issued for the International Year of the Child, illustrating the teaching of the geometry of a circle;

- a Russian stamp, celebrating Communist labour teams, showing a group of adult workers studying the trigonometry of a triangle;

- a stamp from Swaziland showing primary school children experimenting with balances;

- a stamp from the British Virgin Islands showing a class learning to use a computer;

- a stamp from St Lucia, issued for International Literacy Year, showing the teaching of the names of numbers;

- a stamp from Colombia depicting a children's counting frame (a simple form of abacus), widely used for teaching children to count and perform simple arithmetical calculations;

- a stamp from the Maldive Islands, issued for International Education Year, illustrating the use of television to teach a class of children the geometry of a triangle.

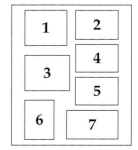

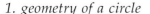

1. geometry of a circle   *2. adults studying trigonometry*
3. children with computer   *4. schoolchildren with balances*
5 teaching names of numbers   *6. counting frame*
7. teaching by television

Metrication

Throughout the centuries various counting systems have been used for weights and measures. As we have seen, the Babylonians (c.1800 BC) used a number system based on 60, while the Mayans of 300–1000 had a mixed calendar system based on 18 and 20. Native American tribes used a variety of number systems based on 5, 10 and 20, and until 1971 the British monetary system had 12 pennies in a shilling and 20 shillings in a pound. The binary system, based entirely on 0 and 1, is used extensively in computer science.

Shortly after the French Revolution, a commission was set up to standardise the weights and measures in France and introduce a metric system. The chairman of this commission was Joseph-Louis Lagrange and its members included Pierre-Simon Laplace and Gaspard Monge (see pages 66 and 68). Thomas Jefferson (see page 64), while ambassador in Paris, was impressed by these events, but it was not until 1866 that the United States Congress passed a law legalising the use of metric measurements.

Since then, most countries have adopted the metric system and in 1960 the international SI system ('Système international') of units was introduced; the metre was standardised as 1,650,763.73 wavelengths of the orange-red light from krypton-86 gas in a vacuum. Even so, some countries, including Britain and the United States, continue to use archaic systems of weights and measures.

A substantial number of metrication stamps have appeared—for example:

- an allegorical figure representing the French metric system;

- a Brazilian metric ruler;

- a Rumanian stamp demonstrating that a metre is one ten-millionth of the distance from the north pole to the equator;

- a stamp from Pakistan demonstrating the metric units of weight, capacity and length;

- a Ghanaian stamp indicating that a metre of cloth is a little more than 3 feet 3 inches;

- two Australian cartoon stamps featuring the metric conversion of length and temperature.

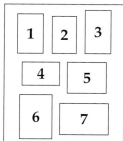

1. *French metric system*　　2. *length*
3. *metric ruler*　　　　　　4. *temperature*
5. *metre*　　　　　　　　　6. *metric units*
7. *metric length*

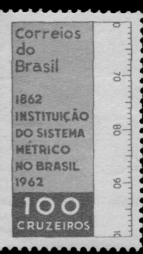

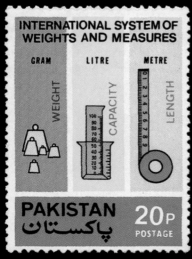

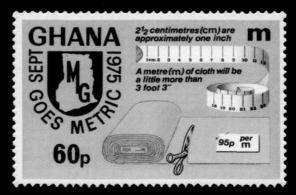

Mathematical Shapes

Everyone is familiar with the usual rectangular stamps issued by all countries. However, several countries have produced stamps whose shapes are more unusual. Such stamps may be issued for decorative reasons, or may be designed to represent a particular object, such as a rugby ball, a map of the country or a lunar module. There are even stamps from Tonga in the shape of a banana, a watermelon and a coconut.

Many countries have issued triangular stamps. The triangles may be equilateral (either way up) or isosceles right-angled (again, either way up). Examples of scalene triangles, in which all of the angles and sides are different, are rare; an early stamp of this type is the scalene triangle issued by Colombia in 1869.

Although four-sided stamps are normally rectangular, they can also take the form of other parallelograms or rhombuses. There are also trapezium-shaped stamps of various kinds; an unusual set from Malaysia consists of stamps shaped like a lunar module.

A few countries have issued pentagonal stamps. Indonesia and the United States have produced stamps whose shape is a regular pentagon, while Malta issued a set of Christmas stamps in the form of an irregular pentagon, representing the crib at Bethlehem. Hexagonal stamps, in the form of a honeycomb tessellation and depicting various aspects of bee-keeping, were issued by the Pitcairn Islands. An early set of octagonal stamps was produced by Thessaly; these came supplied with four corners that were removed before mailing.

Many countries have issued circular stamps, often to commemorate sporting events. Examples include a golf-ball stamp from the Isle of Man and a French football stamp, and France has also produced a stamp in the shape of a rugby ball. A set of semi-circular stamps has been produced by Singapore and a set of elliptical stamps was issued by Sierra Leone.

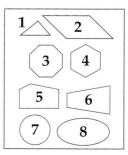

1. *scalene triangle*	2. *parallelogram*
3. *regular octagon*	4. *regular hexagon*
5. *irregular pentagon*	6. *lunar module trapezium*
7. *circular football*	8. *ellipse*

List of Stamps

This list contains all the stamps featured in the book, together with their country of origin, date of issue, and number in the Stanley Gibbons stamp catalogue. We thank Stanley Gibbons Ltd. for permission to use their copyright numbers.

Frontispiece

Dürer's 'Melencolia I' (Mongolia 1978)

Title page

Ernest Rutherford (Russia 1971: SG 3973)

Page viii:

Ten mathematical formulae (Nicaragua 1971: SG 1763–72)

Pages 2–3: Early Mathematics

1. finger counting (Mexico 1939: SG 641)
2. finger counting (Iran 1966: SG 1471)
3. geometrical cave art (Bolivia 1993; SG 1294)
4. Sumerian accounting tablet (Venda 1982, SG 61)
5. Babylonian tablet and comet (Bhutan 1986, SG 669)
6. Stonehenge (Gambia 1997: SG 2703)
7. Carnac (France 1965: SG 1688)

Pages 4–5: Egypt

1. King Djoser's pyramid (Egypt 1987: SG 1649)
2. Gizeh and pyramids (Hungary 1980: SG 3307)
3. pyramids of Gizeh (Congo 1978: SG 627)
4. Egyptian accountants (Egypt 1969: SG 1037)
5. Egyptian papyrus (East Germany 1981: SG E2348)
6. Imhotep (Egypt 1928: SG 176)
7. Thoth (Egypt 1925: SG 125)

Pages 6–7: Greek Geometry

1. $3^2 + 4^2 = 5^2$ (Greece 1955: SG 743)
2. Greek coin showing Pythagoras (Greece 1955: SG 742)
3. Pythagoras' theorem (Macedonia 1998: SG 198)

4. Pythagoras' theorem (Nicaragua 1971: SG 1769)
5. Thales of Miletus (Greece 1994: SG 1947)
6. Pythagoras ('School of Athens') (Sierra Leone 1983: SG 738)
7. Democritus (Greece 1961: SG 876)

Pages 8–9: Plato's Academy

1. Parthenon (Gabon 1978: SG 674)
2. Roman statue of Aristotle (Uruguay 1996: SG 2288)
3. bust of Plato (Greece 1998: SG 2086)
4. Plato and Aristotle ('School of Athens') (Greece 1978: SG 1420)
5. Byzantine fresco of Aristotle (Greece 1978: SG 1422)
6. Greek map and base of statue (Greece 1978: SG 1421)

Pages 10–11: Euclid and Archimedes

1. Alexandria (Hungary 1980: SG 3306)
2. Euclid (Maldive Islands 1988: SG 1260)
3. Archimedes (Greece 1983: SG 1618)
4. Euclid and his pupils ('School of Athens') (Sierra Leone 1983: SG 737)
5. Archimedes (Ribera portrait) (Spain 1963: SG 1559)
6. Archimedes and screw (Italy 1983: SG 1801)

Pages 12–13: Greek Astronomy

1. Ptolemaic planetary system (Burundi 1973: SG 824–7)
2. Hipparchus (Greece 1965: SG 994)
3. Aristarchus' planetary system (Greece 1980: SG 1513)
4. Aristarchus' theory and diagram (Greece 1980: SG 1512)
5. Eudoxus' solar system (Liberia 1973: SG 1177)

Pages 14–15: Mathematical Recreations

1. senet board (Egypt 1965: SG 846)
2. morabaraba (Botswana 1977: SG 403)

3. bhagchal board (Nepal 1974: SG 302)
4. playing eklan (Ivory Coast 1984: SG 811)
5. dakon (tjongkak) (Indonesia 1954: SG 678)
6. eklan board (Ivory Coast 1984: SG 810)
7. Cretan labyrinth (Greece 1963: SG 915)

Pages 16–17: China

1. Zu Changzhi (China 1955: SG 1661)
2. Liu Hui's evaluation of π (Micronesia 1999: SG 809)
3. Zhang Heng (China 1955: SG 1660)
4. distance-measuring cart (China 1953: SG 1603)
5. armillary sphere (China 1953: SG 1604)
6. Chinese abacus (Liberia 1999)
7. Guo Shoujing (China 1962: SG 2062)
8. arithmetical triangle (Liberia 1999)

Pages 18–19: India

1. Vedic manuscript (Mauritius 1980: SG 602)
2. Ashoka column capital (India 1947: SG 301)
3. Indian chess piece (Vietnam 1983: SG 583)
4. Ashoka column, Lumbini (Nepal 1996: SG 644)
5. Aryabhata satellite (India 1975: SG 762)
6. Jantar Mantar (Jaipur 1947: SG 74)

Pages 20–21: Mayans and Incas

1. Mayan city of Tikal (Guatemala) (Bhutan 1998: SG 1190)
2. Mayan calendar stone (Guatemala 1939: SG 400)
3. Mayan observatory (Mexico) (Mexico 1969: SG 1196)
4. Dresden codex (Mexico 1971: SG 1223)
5. Inca messenger (Rwanda 1974: SG 620)
6. Peruvian quipu (Peru 1972: SG 1148)
7. Dresden codex (East Germany 1981: SG E2349)

Pages 22–23: Early Islamic Mathematics

1. Arabic science (Tunisia 1980: SG 951)
2. al-Biruni (Pakistan 1973: SG 363)
3. al-Khwarizmi (Russia 1983: SG 5359)
4. ibn Sinah (Avicenna) (Qatar 1971: SG 348)
5. al-Kindi (Syria 1994: SG 1903)
6. al-Farabi (Turkey 1950: SG 1439)

Pages 24–25: The Middle Islamic Period

1. al-Haitham's optics (Pakistan 1969: SG 286)
2. Omar Khayyam: 'myself when young' (Dubai 1967: SG 249)
3. Nasir al-Din al-Tusi (Iran 1956: SG 1088)
4. Istanbul astronomers (Ascension 1971: SG 136)
5. Persian planisphere (Iran 1956: SG 1087)
6. Omar Khayyam (Albania 1997: SG 2652)

Pages 26–27: Late Islamic Mathematics

1. Córdoba Mezquita (Spain 1986: SG 2891)
2. al-Zarqali and astrolabe (Spain 1986: SG 2893)
3. Maimonides (Spain 1967: SG 1851)
4. Arabic tile (Portugal 1981: SG 1830)
5. ibn Rushd (Averroës) (Lesotho 1999: SG 1658)
6. al-Kashani (Iran 1979: SG 2135)
7. Ulugh Beg's observatory (Turkey 1983: SG 2810)

Pages 28–29: The Middle Ages

1. Ramon Lull (Spain 1963: SG 1597)
2. Nicholas of Cusa (Transkei 1984: SG 160)
3. Gerbert of Aurillac (France 1964: SG 1650)
4. Albertus Magnus (Germany 1980: SG 1927)
5. Albertus Magnus (Belgium 1969: SG 2104)
6. Geoffrey Chaucer (St Vincent 1990: SG 1596)
7. clock of Imms (Austria 1984: SG 2034)

Pages 30–31: The Growth of Learning

1. University of Bologna (Italy 1988: SG 2001)
2. university scholars (Germany 1957: SG 1182)
3. printing press (Finland 1942: SG 381)
4. Bologna students (Guyana 2000)
5. arithmetic and geometry (Netherlands Antilles, 1966: SG 480)
6. astronomy and music (Netherlands Antilles, 1966: SG 481)
7. Adam Riese (Germany 1959: SG 1225)
8. Luca Pacioli (Italy 1994: SG 2247)

Pages 32–33: Renaissance Art

1. Dürer's 'St Jerome in his study' (Panama 1967: SG 954)
2. Roman mosaic (Albania 1969: SG 1359)
3. Leonardo da Vinci (Monaco 1969: SG 953)
4. Filipo Brunelleschi (Italy 1977: SG 1518)
5. Leon Battista Alberti (Italy 1972: SG 1333)
6. Piero della Francesca's 'Madonna and child with saints' (San Marino 1992: SG 1441–3)

Pages 34–35: Go and Chess

1. Arabs playing chess (Mali 1999)
2. Seven-year-old playing Go (North Korea 1997)
3. royal chess party (Djibouti 1980: SG 791)
4. Go formations (China 1993: SG 3842)
5. Moorish chess in Spain (Yemen 1967: SG R346)
6. Caxton's 'Game of chesse' (Great Britain 1976: SG 1016)

Pages 36–37: The Age of Exploration

1. Prince Henry the Navigator (Portugal 1994: SG 2365)
2. Christopher Columbus (Chile 1992: SG 1471)
3. Vasco da Gama (Portugal 1969: SG 1374)
4. Ferdinand Magellan (Bequia 1988)
5. mariner with astrolabe (Portugal 1989: SG 2128)

Pages 38–39: Map Making

1. Abraham Ortelius (Belgium 1942: SG 992)
2. Gerard Mercator (Belgium 1962: SG 1813)
3. Gerard Mercator (Belgium 1994: SG 3227)
4. Mercator projection (Canada 1898: SG 168)
5. Pedro Nunes (Portugal 1978: SG 1723)
6. Ptolemy edition, 1540 (Bophuthatswana 1993: SG 297)
7. Mercator map (Bophuthatswana 1993: SG 299)
8. Ortelius map (Cuba 1973: SG 2083)

Pages 40–41: Globes

1. Arabian celestial globe (East Germany 1972: SG E1510)
2. terrestrial globe (East Germany 1972: SG E1511)
3. globe clock (East Germany 1972: SG E1512)
4. heraldic celestial globe (East Germany 1972: SG E1515)

Pages 42–43: Navigational Instruments

1. armillary sphere (Austria 1966: SG 1481)
2. mariner's astrolabe (St Christopher, Nevis, Anguilla 1970: SG 216)
3. sextant (St Helena 1977: SG 336)
4. quadrant (Turkey 1961: SG 1944)
5. Drake's astrolabe (British Virgin Is 1997: SG 985)
6. Jacob's staff (Netherlands 1986: SG 1483)
7. octant (Germany 1981: SG 1957)
8. back-staff (Portugal 1993: SG 2321)

Pages 44–45: Nicolaus Copernicus

1. Copernicus with planetary system and 'De Revolutionibus' (Venezuela 1973: SG 2203-5)
2. Giordano Bruno (Bulgaria 1998)
3. Copernicus portrait (Matejko) (Pakistan 1973: SG 341)
4. Copernicus with title page and heliocentric diagram (East Germany 1973: SG E1562)

Pages 46–47: The New Astronomy

1. Kepler and planetary system (Hungary 1980: SG 3348)
2. Tycho Brahe at Uraniborg (Ascension 1971: SG 137)
3. Galileo teaching at Padua (Italy 1942: SG 575)
4. Tycho Brahe (Denmark 1946: SG 349)
5. Kepler's first two laws (Germany 1971: SG 1594)
6. Galileo's drawing of the moon (Ascension 1971: SG 138)
7. Galileo Galilei (Russia 1964: SG 2989)

Pages 48–49: Calendars

1. Gregorian calendar (Germany 1982: SG 2009)

2. Gregory XIII with edict (Vatican City 1982: SG 788)
3. Julius Caesar (Italy 1945: SG 640)
4. Greenwich meridian (Great Britain 1984: SG 1254)
5. Greenwich observatory (Great Britain 1975: SG 977)
6. international date line (Tonga 1984: SG 888)
7. reaching the millennium (Great Britain 1999: SG 2069)

Pages 50–51: Calculating Numbers

1. Simon Stevin (Belgium 1942: SG 988)
2. arithmetical symbols (Colombia 1968: SG 1233)
3. Johan de Witt (Netherlands 1947: SG 658)
4. Napier's logarithms (Nicaragua 1971: SG 1768)
5. slide rule (Romania 1957: SG 2504)
6. Jurij Vega (Yugoslavia 1955: SG 785)
7. Jurij Vega (Slovenia 1994: SG 232)

Pages 52–53: Seventeenth-Century France

1. Descartes and optics diagram (Monaco 1996: SG 2282)
2. 'Discours' (incorrect title) (France 1937: SG 574)
3. 'Discours' (correct title) (France 1937: SG 575)
4. Blaise Pascal (Monaco 1973: SG 1079)
5. folium of Descartes (Albania 1996: SG 2639)
6. Blaise Pascal (France 1962: SG 1576)
7. Fermat's last theorem (Czech Republic 2000: SG 267)

Pages 54–55: Isaac Newton

1. apple and *Principia* title (Great Britain 1987: SG 1351)
2. Isaac Newton (Vietnam 1986: SG 913)
3. elliptical planetary motion (Great Britain 1987: SG 1352)
4. Newton's gravitation (Monaco 1987: SG 1847)
5. Newton and diagram (North Korea 1993: SG N3338)
6. binomial theorem (North Korea 1993: SG N3337)
7. law of gravitation (Nicaragua 1971: SG 1764)

Pages 56–57: Reactions to Newton

1. Voltaire (Dubai 1971: SG384)
2. Voltaire (Monaco 1994: SG 2198)
3. la Condamine's mission (Ecuador 1936: SG 530)
4. Jorge Juan (Spain 1974: SG 2240)
5. Maupertuis' mission (Finland 1986: SG 1108)
6. Bishop Berkeley (Ireland 1985: SG 620)

Pages 58–59: Continental Mathematics

1. Leibniz and diagram (Germany 1996, SG 2719)
2. Christiaan Huygens (Netherlands 1929: SG 376a)
3. Leibniz in Hannover (St Vincent 1991: SG 1758)
4. Bernoulli's law of large numbers (Switzerland 1994: SG 1281)
5. Euler in Russia (Russia 1957: SG 2070)
6. Euler and $e^{i\varphi} = \cos \varphi + i \sin \varphi$ (Switzerland 1957: SG J167)
7. Leonhard Euler (East Germany 1950: SG E20)
8. Euler's polyhedron formula (East Germany 1983: SG E2542)

Pages 60–61: Halley's Comet

1. Edmond Halley and map (Mauritius 1986: SG 720)
2. caricature of Halley as comet (Great Britain 1986: SG 1312)
3. Halley and Greenwich observatory (Ascension 1986: SG 394)
4. 1066 comet on Bayeux tapestry (Montserrat 1986: SG 682)
5. 1301 comet on Giotto painting (Montserrat 1986: SG 683)
6. planisphere of the southern stars (St Helena 1986: SG 484)

Pages 62–63: Longitude

1. James Cook and sextant (New Zealand 1997: SG 2051)
2. Cook and transit of Venus (Tuvalu 1979: SG 125)
3. Jean-Dominique Cassini (St Pierre and Miquelon 1968: SG 450)
4. Kendall's chronometer (Ascension 1979: SG 243)

5. Louis de Bougainville (France 1988: SG 2819)
6. Harrison's H1 chronometer (Ascension 1971: SG 140)
7. Harrison's H4 chronometer (Great Britain 1993: SG 1656)

Pages 64–65: The New World

1. Benjamin Franklin (Great Britain 1976: SG 1005)
2. lightning experiment (USA 1956: SG 1075)
3. Benjamin Banneker (USA 1980: SG 1787)
4. Virginia rotunda (USA 1979: SG 1754)
5. Banneker and Washington (Turks & Caicos Is 1982: SG 701)
6. James Garfield (USA 1986)
7. Navajo blanket (USA 1986: SG 2231)
8. Thomas Jefferson (Micronesia 1993, SG 313)

Pages 66–67: France and the Enlightenment

1. Fontanelle and the Académie des Sciences (France 1966: SG 1721)
2. Jean D'Alembert (France 1959, SG 1430)
3. Comte de Buffon (France 1949: SG 1084)
4. Diderot's *Encyclopédie* (Wallis and Futuna Is 1984: SG 448)
5. Marquis de Condorcet (France 1989)
6. Pierre-Simon Laplace (France 1955: SG 1257)
7. Joseph Louis Lagrange (France 1958, SG 1371)

Pages 68–69: The French Revolution

1. Napoleon Bonaparte (France 1972, SG 1976)
2. Napoleon in Egypt (France 1972, SG 1977)
3. Gaspard Monge (France 1990)
4. Gaspard Monge (France 1953: SG 1175)
5. Lazare Carnot (France 1950: SG 1097)
6. Augustin-Louis Cauchy (France 1989: SG 2903)

Pages 70–71: The Liberation of Geometry

1. Gauss and 17-sided polygon (East Germany 1977: SG E1930)
2. Carl Friedrich Gauss (Germany 1955: SG 1130)

3. Gaussian (complex) number plane (Germany 1977: SG 1818)
4. Gauss and Göttingen (Nicaragua 1994: SG 3350)
5. Farkas Bolyai (Hungary 1975: SG 2942)
6. Nikolai Lobachevsky (Russia 1951: SG 1710)
7. János Bolyai (Romania 1960: SG 2765)

Pages 72–73: The Liberation of Algebra

1. William Rowan Hamilton (Ireland 1943: SG 131)
2. Niels Henrik Abel (Norway 1929: SG 215)
3. Vigelund statue (Norway 1983: SG 917)
4. Hamilton's quaternions (Ireland 1983: SG 557)
5. Charles Dodgson (Mali 1982: SG 905)
6. Évariste Galois (France 1984: SG 2607)
7. Richard Dedekind (East Germany 1981: SG E2318)
8. Dunsink Observatory (Ireland 1985: SG 605)

Pages 74–75: Statistics

1. Florence Nightingale (British Virgin Is 1983: SG 504)
2. Amenhotep (Egypt 1927: SG 174)
3. Adolphe Quetelet (Belgium 1974: SG 2376)
4. population graph (Germany 1989: SG 2282)
5. population graph (Austria 1979: SG 1838)
6. gross national product (Norway 1976: SG 762)
7. best-fit curve (Australia 1974: SG 584)

Pages 76–77: China and Japan

1. Matteo Ricci (China–Taiwan 1983: SG 1483)
2. Japanese abacus (Japan 1987: SG 1915)
3. Takakazu Seki (Japan 1992: SG 2208)
4. Xu Guangqi (China 1980: SG3021)
5. Hua Loo-Keng (China 1988: SG 3552)
6. Goldbach's conjecture (China 1999: SG 4448)
7. Japanese shogi (Mali 1999)

Pages 78–79: Russia

1. Alexandr Lyapunov (Russia 1957: SG 2088)

2. Mikhail Ostrogradsky (Russia 1951: SG 1739)
3. Pafnuty Chebyshev (Russia 1946: SG 1187)
4. Konstantin Tsiolkovsky (Russia 1986: SG 5639)
5. Sonya Kovalevskaya (Russia 1996: SG 6595)
6. Otto Schmidt and polar camp (Russia 1933: SG 679)
7. Nikolai Zhukovsky (Russia 1963: SG 2888)

Pages 80–81: Eastern Europe

1. Ruđer Bošković (Yugoslavia 1987: SG 2359)
2. Hermann Oberth (Romania 1989: SG 5261)
3. Ion Ionescu (Romania 1995: SG 5770)
4. Ruđer Bošković (Croatia 1943: SG 121)
5. Bernardo Bolzano (Czechoslovakia 1981: SG 2568)
6. Czech mathematical society (Czechoslovakia 1987: SG 2887)
7. Enigma code-breakers (Poland 1983: SG 2889)
8. *Gazeta matematica* (Romania 1945: SG 1760)

Pages 82–83: Mathematical Physics

1. Friedrich Bessel (Nicaragua 1994: SG 3351)
2. Friedrich Bessel (Germany 1984: SG 2067)
3. Jean Foucault (France 1958: SG 1373)
4. André-Marie Ampère (France 1936: SG 543)
5. François Arago (France 1986: SG 2704)
6. Hermann von Helmholtz (Germany 1994: SG 2594)
7. Foucault's pendulum (France 1994: SG 3225)

Pages 84–85: The Nature of Light

1. Maxwell's equations (Nicaragua 1971: SG 1767)
2. James Clerk Maxwell (San Marino 1991: SG 1410)
3. Heinrich Hertz (Germany 1994: SG 2557)

4. photoelectric effect (Germany 1979: SG 1900)
5. Hertzian radio waves (Portugal 1974: SG 1533)
6. Hendrik Lorentz (Netherlands 1929: SG 375a)
7. Max Planck (Sweden 1978: SG 989)
8. Planck's constant (East Germany 1958: SG E363)

Pages 86–87: Einstein's Theory of Relativity

1. Albert Einstein (Monaco 1979: SG 1412)
2. Albert Einstein (Italy 1979: SG 1595)
3. Einstein with violin (Togo 1979: SG 1354)
4. Einstein quotation (Gibraltar 1998: SG 847)
5. Einstein's law (Nicaragua 1971: SG 1765)
6. Einstein with equation (Ireland 2000: SG 1303)

Pages 88–89: Quantum Theory

1. Niels and Margrethe Bohr (Denmark 1985: SG 809)
2. Louis de Broglie (Sweden 1982: SG 1136)
3. Werner Heisenberg (Sweden 1982: SG 1138)
4. Paul Dirac (Sweden 1982: SG 1137)
5. Paul Dirac (Guyana 1995: SG 4550)
6. de Broglie's law (France 1994: SG 3201)
7. Erwin Schrödinger (Sweden 1982: SG 1135)

Pages 90–91: The Twentieth Century

1. Bertrand Russell (India 1972: SG 667)
2. Srinavasa Ramanujan (India 1962: SG 463)
3. Julia set fractal (Israel 1997: SG 1383)
4. Norbert Wiener (Israel 1999: SG 1439)
5. King Oscar II (Sweden 1891: SG 45c)
6. Henri Poincaré (France 1952: SG 1154)
7. Constantin Carathéodory (Greece 1994: SG 1948)
8. Stefan Banach (Poland 1982: SG 2852)

Pages 92–93: The Birth of Computing

1. Schickard's machine (Germany 1973: SG 1670)
2. Joseph Marie Jacquard (France 1934: SG 520)
3. Charles Babbage (Great Britain 1991: SG 1547)

4. punched tape (Switzerland 1970: SG 786)
5. ENIAC computer (Marshall Is 1999: SG 1075)
6. punched card (Norway 1969: SG 640)
7. computer operator (Ivory Coast 1972: SG 393)
8. John von Neumann (Hungary 1992: SG 4106)

Pages 94–95: The Development of Computing

1. visual display unit (Ireland 1985: SG 623)
2. King Bhumibol at e-mail (Thailand 1997: SG 1942)
3. isometric projection (Netherlands 1970: SG 1106)
4. World Wide Web (United States 2000: SG 3776)
5. computer drawing (Hungary 1988: SG 3843)
6. integrated circuit (Japan 1980: SG 1582)
7. Tim Berners-Lee (Marshall Is 2000: SG 1306)

Pages 96–97: The International Scene

1. Chicago 1893 (USA 1893: SG 236)
2. Moscow 1966 (Russia 1966: SG 3244)
3. Helsinki 1978 (Finland 1978: SG 936)
4. Warsaw 1982/3 (Poland 1982: SG 2850)
5. Kyoto 1990 (Japan 1990: SG 2109)
6. Berlin 1998 (Germany 1998: SG 2862)
7. Budapest 1996 (Hungary 1996: SG 4300)
8. World Mathematical Year 2000 (Luxembourg 2000: SG 1522)

Pages 98–99: Mathematics and Nature

1. logarithmic spiral (Switzerland 1987: S1121)
2. snowflake (Bulgaria 1970: SG 2052)
3. pine cone (Israel 1961: SG 220)
4. ammonite (Hungary 1969: SG 2467)
5. sunflower (Great Britain 1996: SG 1961)
6. fluorite crystals (Switzerland 1961: SG 649)
7. sulphide crystals (Germany 1968: SG 1452)
8. honeycomb (Luxembourg 1973: SG 908)

Pages 100–101: Twentieth-Century Painting

1. Vasarely's *'Tetcye'* (France 1977: SG 2176)
2. Delaunay's *'Rhythme: Joie de vivre'* (France 1976: SG 2110)

3. Vasarely's *'Vega-Chess'* (Hungary 1979: SG 3273)
4. Mondrian's *'Broadway Boogie-Woogie'* (Liberia 1997)

Pages 102–103: The Geometry of Space

1. impossible triangle (Sweden 1982: SG 1105)
2. Escher's impossible cube (Austria 1981: SG 1908)
3. impossible object (Sweden 1982: SG 1106)
4. sculpture 'Continuity' (Switzerland 1974: SG 880)
5. Escher posthorn design (Netherlands 1949: SG 706)
6. Escher with mural (Netherlands 1998: SG 1891)
7. sculpture 'Expansion' (Brazil 1953: SG 843)
8. German pavilion (Germany 1997)
9. Möbius strip (Brazil 1967: SG 1180)

Pages 104–105: Mathematical Recreations

1. noughts-and-crosses (Netherlands 1973: SG 1180)
2. maze (Netherlands 1973: SG 1181)
3. dominoes (Netherlands 1973: SG 1182)
4. Rubik cube (Hungary 1982: SG 3449)
5. computer chess (Israel 1990: SG 1124)
6. fox-and-geese (Sweden 1985: SG 1267)
7. bridge (Monaco 1976: SG 1240)
8. Chinese checkers (Sweden 1985: SG 1270)

Pages 106–107: Mathematics Education

1. geometry of a circle (Guinea-Bissau 1980: SG 616)
2. adults studying trigonometry (Russia 1961: SG 2633)
3. children with computer (British Virgin Is 1996: SG 934)
4. schoolchildren with balances (Swaziland 1984: SG 447)
5. teaching names of numbers (St Lucia 1990: SG 1051)
6. counting frame (Colombia 1977: SG 1433)
7. teaching by television (Maldive Is 1970: SG 352)

Pages 108–109: Metrication

1. French metric system (France 1954: SG 1224)
2. length (Australia 1973: SG 532)
3. metric ruler (Brazil 1962: SG 1062)
4. temperature (Australia 1973: SG 535)
5. metre (Romania 1966: SG 3405)
6. metric units (Pakistan 1974: SG 370)
7. metre length (Ghana 1976: SG 762)

Pages 110–111: Mathematical shapes

1. scalene triangle (Colombia 1869: SG 58)
2. parallelogram (Pakistan 1976: SG 430)
3. regular octagon (Thessaly 1898: SG M165)
4. regular hexagon (Pitcairn Is 1999)
5. irregular pentagon (Malta 1968: SG 409)
6. lunar module trapezium (Malaysia 1970: SG 61)
7. circular football (France 1998: SG 3472)
8. ellipse (Sierra Leone 1969: SG 479)

Bibliography

The following books were found to be particularly useful during the writing of this book.

History of Mathematics

David Burton, *The history of mathematics: an introduction* (Third edition), William Brown, 1991.

Howard Eves, *An introduction to the history of mathematics* (Sixth edition), Saunders, 1990.

John Fauvel and Jeremy Gray (eds.), *The history of mathematics: a reader*, Macmillan, 1987.

Charles C. Gillespie (ed.), *Dictionary of scientific biography* (18 volumes), Scribner, 1970–1990.

Ivor Grattan-Guinness (ed.), *Companion encyclopaedia of the history and philosophy of the mathematical sciences* (2 volumes), Routledge, 1994.

Michael Hoskin (ed.), *The Cambridge illustrated history of astronomy*, Cambridge University Press, 1997.

Victor Katz, *A history of mathematics: an introduction* (Second edition), Addison-Wesley, 1998.

Dava Sobel and William J. H. Andrewes, *The illustrated Longitude*, Fourth Estate, 1998.

Dirk J. Struik, *A concise history of mathematics* (Fourth revised edition), Dover Publications, 1987.

General history

Jerome Burne (ed.), *Chronicle of the World*, Longman and Chronicle Communications, 1989.

Mathematical stamps

Larry Dodson, *Computers on stamps and stationary*, ATA Handbook 134, American Topical Association, 1998.

William L. Schaaf, *Mathematics and science: an adventure in postage stamps*, National Council of teachers of mathematics, 1978.

Peter Schreiber, *Die Mathematik und ihre Geschichte im Spiegel der Philatelie*, Teubner, 1980.

Robin J. Wilson, Various 'Stamp Corners' from *The Mathematical Intelligencer*, Springer, 1984–2000.

Hans Wussing and Horst Remane, *Wissenschaftsgeschichte en miniature*, Deutscher Verlag der Wissenschaften, 1989.

James S. Young, *Collect chess on stamps* (Second edition), Stanley Gibbons Ltd., 1999.

Various articles from *Philamath*, Mathematical Study Unit of the American Topical Association and the American Philatelic Society, 1979–2000.

Acknowledgements

Every effort has been made to obtain permission for the reproduction of the stamps featured in this book. Our thanks are due to the postal authorities of all the countries represented, and in particular to:

Anguilla [General Post Office]; Australia [National Philatelic Collection, Australia Post]; Austria [Post & Telekom Austria]; Belgium [Postage Stamps and Philatelic Department]; Bhutan [Bhutan Post]; Bolivia [Departamento Filatelia, Empresa de Correos de Bolivia]; Canada [Canada Post Corporation]; Czech Republic [Czech Post Philatelic Service (Postfila)]; Denmark [Post Danmark]; Finland [Philatelic Centre, Finland Post Ltd.]; France [La Poste, Service National des Timbres-poste et de la Philatelie], Germany [Deutschepost]; Ghana [Ghana Postal Services Corporation]; Greece [Philatelic Service, Hellenic Post]; Guyana [Post Office Corporation]; Ireland [An Post]; Israel [Israel Postal Authority]; Italy [Macro Divisione Servizi Postali, Poste Italiane]; Malta [Maltapost plc]; Marshall Islands [Special Markets]; Mauritius [Postal Services]; Mexico [Servicio Postal Mexicano]; Monaco [Office des Emissions de Timbres-Poste]; Montserrat [Montserrat Philatelic Bureau Ltd.]; Netherlands [PTT Post]; Netherlands Antilles [Post Netherlands Antilles, Ltd.]; Panama [Departamento de Filatelia y Museo Postal]; Pitcairn Islands; Poland [Poczta Polska]; Portugal [CTT—Correios de Portugal]; Republic of Maldives [Philatelic Bureau]; Romania [Romfilatelia]; St Helena [Post Office Department]; St Pierre et Miquelon [Agence Régionale du Tourisme]; San Marino [Azienda Autonoma di Stato Filatelica e Numismatica]; Slovenia [Po ta Slovenije d.o.o.]; South Africa [South African Post Office Ltd.]; Spain [Comision de Programacion de Emisiones de Sellos y Demas Signos de Franquero]; Sweden [Sweden Post Stamps]; Switzerland [Swiss Post]; Taiwan, Republic of China [Directorate General of Posts]; Thailand [Philatelic Department, The Communications Authority of Thailand]; Tunisia [Centre Directeur de la Philatelie]; Turkey [General Directorate of Post]; Turks and Caicos Islands [Department of Post Office, Ministry of Tourism, Communication and Transportation]; Tuvalu [Tuvalu Philatelic Bureau]; United Kingdom [The Post Office, All Rights Reserved]; United States of America; Vatican City [Ufficio Filatelico e Numismatico]; Venezuela [Instituto Postal Telegrafico de Venezuela]; Vietnam [Vietnam Stamp Company].

We also thank the following organisations:

Stanley Gibbons Publications (for use of their copyright catalogue numbers, from *Stamps of the World*); The Escher Foundation, M.C. Escher/Cordon Art, Baarn-Holland, The Netherlands; ADAGP Paris, 2000 (for permission to use stamps engraved by Albert Decaris and Pierre Gandon);

and the following individuals:

Sven Bang; Pierre Béquet; Louis Briat; H. Chylinski; family of Jacques Comber; Hatim Elmeki; Jaques Gauthier; Marie-Noëlle Goffin; René Guillivic; Jacques Jubert; Adalbert Pilch; Joseph Rajewicz; Valentin Wurnitsch.

Anyone who feels that their copyright has been infringed is invited to contact the publishers, who will correct the situation at the earliest possible opportunity.

Index